TAINTED MILK

TAINTED MILK

Breastmilk, Feminisms, and the Politics of Environmental Degradation

MAIA BOSWELL-PENC

STATE UNIVERSITY OF NEW YORK PRESS

Published by
STATE UNIVERSITY OF NEW YORK PRESS, ALBANY

© 2006 State University of New York

For information, address the State University of New York Press,
194 Washington Avenue, Suite 305, Albany, NY 12210-2384

Production, Laurie Searl
Marketing, Susan Petrie

Library of Congress Cataloging-in-Publication Data

Boswell-Penc, Maia, 1964–
  Tainted milk : breastmilk, feminisms, and the politics of environmental degradation / Maia Boswell-Penc.
    p. cm.
  Includes bibliographical references and index.
  ISBN 0-7914-6719-8 (hardcover : alk. paper) — ISBN 0-7914-6720-1 (pbk. : alk. paper)
  1. Breastfeeding—Health aspects—United States—History. 2. Breast milk—Contamination—United States—History. 3. Infants—United States—Nutrition—History. 4. Food contamination—Environmental aspects. 5. Environmental degradation—Health aspects. 6. Pollution—Health aspects. I. Title.

RJ216.B67   2006
363.7—dc22   2005015756

ISBN-13: 978-0-7914-6719-0 (hardcover : alk. paper)
ISBN-13: 978-0-7914-6720-6 (pbk. : alk. paper)

10 9 8 7 6 5 4 3 2 1

I dedicate this book to my son, Thatcher, and to my daughter, Amelia, each of whom was born and breastfed while I worked on this book. Your sweet smiles and adorable laughter gave me hope that we can—*will*—make the environment safer for all future generations.

CONTENTS

ACKNOWLEDGMENTS

xi

INTRODUCTION

1

Advocating for Breastfeeding: An Uphill Battle

A New Chapter with Familiar Patterns: Biomonitoring

Infant Feeding under Wraps

Breastmilk's Significance

Environmental Toxicity

Laying Out the Players

CHAPTER ONE

THE EVOLVING NARRATIVE OF INFANT FOOD CONTAMINATION:
HISTORICAL VIGNETTES

17

Part I: Wet Nursing

Contamination Narratives: A Series of Supremacist Gestures

Defining Terms

Addressing the (Colonizing) Dangers of Overgeneralizations

*Wet-Nursing and Semen-Curdled Milk: Assessing the Myth of the
Constitution Too Delicate to Suckle*

Classism, Sexism, Ageism, and Lookism: Assessing Some Trends

*The "Threat of the Dangerous Stranger": Wet-Nursing in
Select Mid-nineteenth-Century American Contexts*

Human Milk Banking

White Supremacy and the Colonization of Black Women's Bodies

Part II: Mathematical Formulas

*The Advent of Scientific Infant Food and
the Era of "Scientific Motherhood"*

The Growing Impact of Advertising

Medicine and Markets

*Colonization and the Search for New Markets:
Exporting Infant Food as the More "Civilized" Choice*

Breastfeeding and Population Debates

*Debates Around the Possibility of HIV Transmission Through Breastmilk:
A New Chapter or an Old Story?*

The True Unborn Victims of Violence

CHAPTER TWO

TOXIC DISCLOSURE: THE GROWING AWARENESS OF ENVIRONMENTALLY
CONTAMINATED BREASTMILK IN THE CONTEXT OF MUCH-NEEDED
BREASTFEEDING ADVOCACY

61

Rachel Carson's Legacy

*The Impact of Changing Rates of Breastfeeding in the United States
on Interest in Environmental Pollutants in Breastmilk*

Burgeoning Environmentalist Attention and Pulling Back

Environmentalists' Reluctance

Children's Health Campaigns: The Power of the Internet

Impacts of the Breastfeeding Advocacy Community

*Developments in Related Campaigns: Movements
to Curb Formula Advertising and Use and to Protect Infant Health*

Breastfeeding and Environmental Advocates: Friends or Foes?

The Role of the Press in Disseminating Stories of Breastmilk Toxicity

The Press on Toxic Fish

*Developments in Related Campaigns: The Environmental
Justice Movement*

*The Sensitive and Unusual Nature of Breastmilk on a Policy Level
and as a Fluid Vital to Life*

CHAPTER THREE

BREAST FETISHIZATION, BREAST CANCER, AND BREAST AUGMENTATION:
THE CURIOUS OMISSIONS OF BREASTFEEDING AND BREASTMILK
CONTAMINATION AS SIGNIFICANT FEMINIST ISSUES

99

A Lack of Feminist Attention

Feminist Silence on Infant Feeding: From Preoccupation with the Sexual Division of Labor
to Fights for Reproductive Freedoms

Structural Impediments

Results of the Lack of Feminist Attention to Infant Feeding

Feminist Attention to Infant Feeding: "The Scientific" versus "the Political"?

Feminists Critiquing (and Embracing) Science

Who's Doing the Feeding? Racism, Classism, and Colonization in Childcare Work

Immigrant Women Doing Childcare

(In)Attention among Feminists to the Environmental Contamination of Breastmilk

What Science Can Offer: The Benefits of Breastmilk

When the "Personal Is the Political": The Ecological Impacts of Formula-Use

Findings on the Environmental Contamination of Breastmilk and
Resulting Health Effects on Women and Children

Feminist Attention to the Environmental Contamination of Breastmilk

Feminists, Environmental Justice Activists, and Breastmilk Toxicity

Structural Problems Impeding Breastfeeding in the United States: Women in
the Workplace and in Environmentally Devastated Communities

CHAPTER FOUR

POLLUTING THE "WATERS" OF THE MOST VULNERABLE: ENVIRONMENTAL
RACISM, ENVIRONMENTAL JUSTICE, AND BREASTMILK CONTAMINATION

137

Toxic Breastmilk as an "Environmental Justice" Issue

The Environmental Justice Movement: A Range of Documented Cases, with
the List Still Growing

Chemical Colonization: Toxicity on the Akwasasne

The Mother's Milk Project

The Continuing Legacy of Chemical Colonization: The Case of Latina Farmworkers

Breastmilk Contamination among U.S. Farmworkers

Buying Organic and Protecting "Our" Children: How We Forget the "Others"

Moving Toward Environmental Justice

Urban Women and Those at Risk due to Occupational or Other Exposures

Approaches to Remediation: Community-based Participatory Action Research

The Watchperson Project in the Greenpoint/Williamsburg Neighborhood of Brooklyn, NY: Bringing Awareness to Subsistence Fishing Hazards

Greater Movement Toward Environmental Justice: The Institute of Health's Vision and the La Duke, Bonnie Raitt, Indigo Girls Consciousness-Raising Team

CONCLUSION: A NEED FOR MORE ATTENTION, AND MORE CAREFUL ATTENTION TO BREASTMILK TOXICITY

167

Assessing Coverage: Two Widely Divergent Approaches

The Center for Children's Health and the Environment—Basically Appropriate, Bold Attention

NOTES

177

APPENDIX A

185

APPENDIX B

187

REFERENCES

189

INDEX

209

ACKNOWLEDGMENTS

I would like to thank many people for helping me to get this book out to press and helping me to manage my life during the process. First, I want to thank Dr. Stanley Penc, my husband, for being a loving supporter and source of stability throughout this process. Only he could remain so calm while watching me struggle, and while finishing medical school and working on a residency—all with a puppy and two babies in the mix. I would also like to thank my children, Thatcher and Amelia, for being such wonderfully good-natured babies; I could never have gotten so much work done otherwise.

I want to thank the Women's Studies Department at the University at Albany for giving me the opportunity to work and teach alongside them, in particular, the Department Chairs who gave me the opportunity to grow: Vivien Ng, Marjorie Pryse, Judi Barlowe, and Christine Bose. I would also like to extend special thanks to Bonnie Spanier, Judi Barlowe, and Molly Eness for helping me to manage that difficult first draft. I also extend thanks to Tom Cohen for helping me to have the opportunity to work on this book.

Next I would like to thank a group of people who supported me in managing my life during this process: my parents George and Veta Boswell who offered support every step of the way; my sisters, nieces, nephews, and brothers-in-law, Randy Brown and Bob Kondelka, Brianna Brown, Kama Koudelka, Hannah, Sarah and Matthew Brown, Bobby, Danny, and Teddy Koudelka; and my in-laws Buddy and Adeline Penc, Tony Penc, and Melissa and Duane Spignardo. I would also like to thank those who offered emotional support: Pat McNamara, Jennifer Moore, Heideh Kabir, Leslie and Valerie Montana, Sandra and Phil Williams, Tim and Judy Trustcott, the staff at Campus Children's Center, Shealeen Meaney and Laura Kate Boyer. Then there are my other supporters: Suzanne Diamond for watching Amelia and offering other help, and the graduate and undergraduate students at the State University of New York at Albany for filling in with Amelia care and listening to me talk about breastmilk. Also, a thanks to Martha Holzeman.

Finally, I would like to thank all those at SUNY Press who offered such expert advice, including Jane Bunker, my Senior Editor, who made the whole experience such a pleasure, Laurie Searl, my Senior Production Editor, and Susan Petrie, my Marketing Editor. Finally I would like to thank all those whose research has been instrumental in allowing me to craft this book, and who offered inspiration, including Winona LaDuke, Linda Layne, Judith Galtry, Penny Van Esterick, Rima Apple, Janet Golden, Gabrielle Palmer, Rachel Carson, and Sandra Steingraber.

INTRODUCTION

ADVOCATING FOR BREASTFEEDING: AN UPHILL BATTLE

In December 2003, a series of commercials were supposed to have aired on national television. The ads dealt with the importance of breastfeeding and the risks associated with choosing not to breastfeed. Supporters said that running them could help bring about healthier babies, healthier people, and a healthier environment and that they could save money on health expenditures as well as save lives. Critics claimed they were unfair to moms who could not breastfeed. The critics won. Just as the ads' producers—the Ad Council—were about to release the images, the US Department of Health and Human Services (USDHHS) abruptly made the decision to withdraw the campaign from the air. The situation surrounding these commercials, including the timing of this episode in conjunction with the release of several new studies about toxic substances found in breastmilk, offers a glimpse into the complex story of breastfeeding in the United States. It illustrates some of the tensions at play, the arguments made by various players who work to advocate and to silence discussions about breastmilk, and the high stakes involved.

The advertisements featured actors portraying pregnant women performing various outrageous stunts that would be dangerous to a fetus. In one, pregnant women skate by, arms entangled, wrestling one another in a roller-derby match; in another, pregnant women balance precariously in the water, log rolling; in a third, a soon-to-be mom grasps the leather handhold of a saddle as she strains to stay atop a mechanical bull. The commercials' producers, aware of the relatively low breastfeeding rates in the United States and seeking to get the attention of pregnant women and the public who play a role in infant feeding choices, were saying, in effect, "You wouldn't risk your baby's health before it's born. Why start after?" ("Milk"). They suggest, in other words, that breastfeeding is the safer choice and that not breastfeeding, despite what the many ads for formula tell you, does come with some risks.

1

While the commercials may have been overblown, their designers had in mind the task of countering a long and continuing history of formula advocacy in the United States. They wanted to grab people's attention, and get them, perhaps for the first time, to really think about the fact that breastfeeding is more than just a choice equal to using formula.

Although a version of the ads was eventually released, in June 2004, many believe that the changes represent a dramatic watering-down of the original. And in June, several sources covered the events and the controversies involved. For example, on June 4th, 2004, *20/20* ran a story on the advertisements, showing the "controversial commercials that are never going to be seen anywhere else on television" ("Milk" 12). The *20/20* producers released "the never-aired commercials" that they had obtained "from a source involved in the campaign," and offered an investigation into the pulling of the ads from the public eye. Dr. Bobbi Phillip of the Boston Medical Center, one of the many pediatricians involved in the campaign, says the focus on the risks of not breastfeeding made sense, given the many documented benefits of breastmilk, including positive effects on immunity, neurological outcomes, and other functions (Lobet 17).[1] And, as Brian Ross, the ABC news interviewer pointed out, the National Institutes of Health released a study in May 2004 that found that babies who are not breastfed had a 20 percent higher risk of death in their first year ("Milk" 12). Dr. Phillip characterized breastfeeding as "a huge health issue, similar to smoking" and reminded us of how outrageous and "unfair" the antismoking ads that, in effect, blame smokers for the deaths of loved ones due to secondhand smoke might have seemed early in the antismoking campaign. She explains that she drew this comparison "because we know that there are significant risks to not breastfeeding" (12). So why did the DHHS, then run by Tommy Thompson, a member of President Bush's cabinet, and the official sponsor of the campaign, choose to censor the ads? John Stossel of ABC News asked if it was because the commercials were "unfair to moms who can't breast feed" or, as chief investigative correspondent Brian Ross wondered, "Was it money talking?"

Beginning his search, Ross asked what role the infant-formula industry played in having the campaign put on hold. Directing his questions to Christina Beato, acting Assistant Secretary of DHHS who was speaking for Secretary Thompson, he receives the answer that "it was the department's decision alone to hold back the spots because of questions about their accuracy" (13). Ross pointed out that the formula industry's trade group hired a well-connected Washington lobbyist, Clayton Yeutter, Secretary of Agriculture for the first President Bush and former Republican party chairman, and that "Yeutter and industry officials were able to get something the breastfeeding advocates could not—a private meeting at the Department of Health and Human Services, with Secretary Thompson, to discuss the breastfeeding campaign" (13). Ross then upped the pressure and asked "Why is it the industry gets to meet with the Secretary, Mr. Thompson, but the breastfeeding advocacy groups do not? They didn't hire the right lobbyists, is that what you're saying?" Beato's response, "I'm telling you, they [presumably the pediatricians and advocates] have been working with this department all along," is followed by two more questions from Ross. First, "Did you meet with the people from the infant-formula companies?," to which Beato responds, "Yes," and, second, "Did you meet with the people from advocacy groups?" Beato's answer, that no, she did not meet with the advocates because "they never asked to meet with me" is baffling, given the passion about the campaign on the part of advocates. Ross then pointed to the seemingly cozy relationship between Yeuter and the DHHS.

While Beato argued that "I took out leukemia and we took out the percentages. There is no science—to back up the percentages," others stand behind the accuracy of the figures used. One of those in support of the figures is Dr. Larry Gartner, former director of pediatrics at the University of Chicago Medical School and now the head of the breastfeeding committee for the American Academy of Pediatrics (AAP). In response to Ross's question whether, "when it comes to the scientific validity of this campaign," he is prepared to support it, Dr. Gartner said that he "absolutely" supports it, and that the campaign "is backed up by scientific research that has been reviewed by two different panels" (15). Noting that the AAP "gets several million a year from formula companies," and that "top leaders met privately with industry officials about the campaign but never told Dr. Gartner," Ross digs deeper. He questioned Dr. Joe Sanders, the executive director of the academy, who claims that "there is no such thing as corporate influence over the American Academy of Pediatrics" (16). When Ross notes that the academy questioned the approach and accuracy of the campaign, Sanders concurs, saying, "We saw the information from the Ad Council. And they indicated that there was something about a woman riding a bull, a pregnant woman riding a mechanical bull. There was a suggestion. . . ." Ross's response "What's wrong with that?" prompts Sanders to say, "Well, I don't think a pregnant woman belongs on a mechanical bull. Do you?," to which Ross asks "Did you see the commercial to see the context of it?" Sanders admits he never saw the ads.

Ultimately a redone advertisement was unveiled, one with such substantial changes that many who support breastfeeding have been outraged. The voice-over in the new campaign says "Recent studies show babies who are breastfed are less likely to develop ear infections, respiratory illness, and diarrhea." This approach, likened to saying "If you breastfeed, your baby will be healthier," is very different from saying "If you don't breastfeed exclusively, your baby will be more likely to develop . . . ." It leaves out all references to statistical findings, which studies have shown, are deemed compelling by U.S. audiences, and leaves out findings that using formula means that your baby will have an increased risk of developing leukemia (up 30%), ear infections (up 20%), and diabetes (up 40%) (Ad Council) (16).

I begin with this story because it is important to explain the ongoing power of the formula industry (a $3-billion-a-year conglomerate), the government, and other institutions in helping to curtail pro-breastfeeding messages while also giving liberal license to the formula industry. As Brian Ross pointed out, "Until now, the infant-formula companies have had the airwaves to themselves, advertising their products as close to mother's milk" (13). While most other industrialized nations, as well as many others (118 in all), have signed the International Labor Organization's (ILO) ban on the direct marketing of breastmilk substitutes to doctors and to women, the United States, bowing to industry pressure, has not. The result? We all see the myriad ads proclaiming how various formulas have properties like breastmilk, yet none of us will see a single advertisement asking us to think about whether there might be risks involved in the choice not to use breastmilk.

Although higher than at most points in recent history, rates of breastfeeding remain low in the United States in comparison to many other countries, with only about 27 percent of women still nursing to some degree at six months after birth, and only 7.9 percent nursing exclusively at six months.[2] At twelve months, 12.3 percent of infants continue to receive at least some breastmilk. The story of the breastfeeding ads tells us much about the long history of breastmilk and I want now to turn to another event in breastfeeding history, one that took place at the same time that the controversy around the advocacy ads was unfolding. This story involves the release of biomonitoring findings, and the publicizing of the results of several studies showing that all U.S. women's breastmilk contains surprisingly high amounts of the toxic chemicals PBDEs , or fire retardants, as well as other substances that persist in the body. The timing of these two events, where inattention to the importance of breastmilk gets sidetracked just as attention to the environmental contamination of breastmilk reaches a new pitch, forms a pattern. Whether accidental or not, again and again events play out in such a way that attention to breastmilk issues gets curtailed. One consequence has been that we in the United States continue to view breastmilk as not all that important. Another has been that we, as a nation, have not woken up to the significance of the findings that women's breastmilk in the United States contains what some characterize as high levels of toxic pollutants.

A NEW CHAPTER WITH FAMILIAR PATTERNS: BIOMONITORING

In 2003, two reports were released, one by the Centers for Disease Control (CDC), the other by the Mount Sinai School of Medicine in collaboration with the Environmental Working Group and Commonweal, that used biomonitoring to document the presence of synthetic chemicals in people's bodies (Sucher and Walker). These studies and several others like them, which measure body burdens, showing the chemical fingerprint, the living proof of what chemicals have built up in our bodies over our lifetime, have become controversial. Breastmilk works well in the studies because it is high in fat, and many persistent, highly toxic chemicals lodge in fatty tissues; it can also be easily extracted. But biomonitoring studies using breastmilk have prompted the question of whether the documenting of chemicals in breastmilk will create a backlash against breastfeeding. Many, including Dr. Ruth Lawrence, who published *Breastfeeding: A Guide for the Medical Profession*, fear that people will not recognize that "This is not an issue of contaminants in breastmilk, but an issue of the fact that the easiest way to measure contaminants is through the use of breastmilk" (Gross-Loh 63). The question has proven much more sensitive in the United States than in other countries, where breastfeeding rates are higher and citizens demand more information about possible toxins in their food and environment (Schmidt 1). I believe this controversy around biomonitoring—one more chapter in a long history which I will discuss in this book—results from the antibreastfeeding climate that continues to haunt the United States, along with the pressure put on advocates to limit discussion of possible toxins in breastmilk.

The Environmental Working Group study, which tested twenty moms in fourteen states, found that all samples contained high levels of PBDEs. A peer-reviewed study published in the November 2003 *Environmental Health Perspectives* analyzed forty-seven samples from U.S. moms and found all "had high to very high levels," according to study leader Arnold Schecter of the University of Texas School of Public Health (Kelly). The CDC study, which is ongoing, tested 116 people, and is finding high levels of many chemicals. We use the retardants throughout our living environments—in car dashboards, computer shells, sofas, foam mattresses, televisions, drapes, furniture, hair dryer casings, carpet padding, insulation, paints, car parts, to name a few—seventy million pounds in the United States and Canada each year (Schmidt). Although no human health studies have been done, animal research has linked these chemicals to thyroid hormone disruption, developmental damage, neurodevelopmental changes (learning and memory deficits), and cancer, among other problems. Scientists focusing on mice and rats given dosages comparable to human exposures, either in utero or early after birth, worry that human exposures are leading to "permanent changes in behavior and learning and memory" (Kelly; Lobet).

Adding to the controversy is the move by California to institute a law that would set up the first statewide collection and analysis of human samples. While the CDC has begun keeping data on what substances are found in the tissue of average Americans, some argue that individuals and communities have a right to know that information. California State Senator Deborah Ortiz, the sponsor of the bill, argues that biomonitoring could become a "powerful political tool because it could reveal geographic differences in exposure" (Lobet). She muses that we could "measure women in East Los

Angeles who live near an incinerator, or women who live in an area in the
Central Valley, where there is a lot of arsenic in water." For her, the logic
is to "get us to a point where we can't turn a blind eye, that we can't turn our
back, to the huge problems and the risks that we are placing on women
throughout California" (16). She hopes that having individual blueprints will
help us arrive there. The studies have prompted environmental biochemist
Kim Hooper and others to ask the question: "whether the current way of
regulating chemicals is sufficiently protective of fetal and infant health"
(15). Hooper points out that "for the last twenty-five years we've been following this paradigm of we need to show a chemical-human disease connection" and it has not worked. Many argue that we need a new paradigm, one
that looks at chemical body burdens and that, like many European countries,
follows the precautionary principle, emphasizing the cost of inaction, rather
than "cost-benefit" analyses (Pohl).

Donne Brownsey, a lobbyist in Sacramento for the Breast Cancer Fund,
supports the California effort, saying it "indicates a shift in public attitude"
(Lobet 16). Indeed, she believes it makes political sense to choose breastmilk,
arguing that "if breastmilk talks, people will listen." As I will suggest in this
book, such a comment does indicate a shift; the climate around breastfeeding
in the United States has been such that it has been extremely dangerous to
point so fully to the environmental contamination of breastmilk, to frame it
as a political tool to raise attention to environmental degradation. But many
opposed to the individualized studies point out that biomonitoring is not
about contaminated breastmilk, but about contaminated bodies, about the
body burdens carried by all. Marian Thompson, one of the founders of La
Leche League, observes: "Environmentalists use breastmilk as a red flag to
gain attention, but the fact of the matter is that in doing so they are making
a lot of women decide not to breastfeed their babies" (Gross-Loh 63). She
thinks breastmilk is used "to create attention about this pure food being
contaminated." Biologist Sandra Steingraber, who has written extensively
on the issue, notes: "The people who are advocating [breastfeeding] in the
public health community, the lactation community, the midwifery community, pediatricians and obstetricians—they're very touchy about any negative comment about breastfeeding and breastmilk" (Lobet 17). University
of California toxicologist Robert Krieger, who supports the kind of federal
biomonitoring conducted by the CDC, does not support the type of effort
being pushed in California. He believes that when individuals receive their
own results, they could be cause for alarm, "wasted money, and an unwar-
ranted fear of chemicals" (18).

Given the antibreastfeeding climate in the United States, advocates'
reluctance to focus on environmental pollutants is understandable. So many
factors contribute to the propensity toward formula-use here, including inadequate maternity leave; inadequate access to lactation specialists; lack of
legislation mandating nursing or pumping breaks and facilities in the work-

place; lack of subsidies for breast pumps; nonsupportive cultural norms; a lack of legislation limiting the direct advertising of formula to women, doctors, and hospitals; and lack of government and other institutional support for breastfeeding. Those of us who think of breastfeeding as *the* best way to nourish our babies—indeed those of us who even consider breastfeeding as an option—have the work of thousands of breastfeeding advocates to thank. While the successes in raising awareness of the importance of nursing can be attributed, in part, to researchers, scientists, and medical personnel conducting research on the relative merits of breastmilk and various formulas, these successes would not have become a reality without the dedication of the breastfeeding advocacy community. The withdrawal of the advertisements reported by *20/20*, along with the collusion of the formula industry, the government, and some in the medical establishment, further solidify the antibreastfeeding climate, thus rendering a program like that proposed in California controversial.

Dana Barr, who has worked on the biomonitoring program at the CDC also expresses reservations about the California proposal, pointing out that "a lot of [the data] aren't that easy to interpret right now on an individual basis" (Lobet 19). She does, however, believe that the geographical testing is "an outstanding idea," offering information relevant to "whether you live close to an agricultural region or whether you live close to an inner city" (19). Her comments express part of the ambivalence. Both those opposed to the pro-breastfeeding ads and those opposed to the breastmilk biomonitoring plan in California argue that this is all for the sake of mothers who might misinterpret, feel afraid, or be uncomfortable. But wouldn't it be better for us to allow mothers to know the facts, hear all sides, and make decisions for themselves? The attitudes suggesting that women need "protecting" and that keep the public at large from having information on which to base healthy, respectful discussions could prevent us from protecting future generations. Feminists have repeatedly suggested the danger of those in power speaking for women, for the "others" who are not in a position to speak. Many of the mothers who did participate in various studies, with results showing that they had high body burdens, said that they would continue to breastfeed, given the properties of breastmilk and the fact that, even when contaminated, it offers benefits, helping to protect against the possible effects of pollutants. And most mothers I have spoken with about the ads aired on *20/20* as well as mothers who were able to breastfeed for only a very short time have said that they would not view such ads as offensive.[3] Steingraber finds the "idea that nursing mothers should be protected against knowledge of what's in their milk . . . profoundly condescending" (Lobet 17). Likening the information to car seat recalls, she suggests that ending the silence and allowing for the sharing of information can benefit us all.

Allowing communities and individuals to receive data under the supervision of educators trained in interpreting the results, as California's bill mandates, would help individuals to feel that they have a personal stake in the issue, and might empower communities to come together to advocate for greater protections.[4] One issue that gets little attention, but is extremely important, concerns environmental injustice and the fact that pollution falls more heavily on certain groups—people of color (particularly Native Americans), people in certain occupations, and people of lower economic means, who live in less protected spaces. For communities and individuals to obtain information about the specific toxins associated with their geographical area would constitute a huge step forward in the fight for environemental justice. So, in a sense, to limit the use of biomonitoring to general averages is to limit the tools that can help bring us toward greater equity and environmental justice. It is to fail to do all we can to protect our citizens and future generations. According to Sharyle Patton, Director of the Health and Environment Program Commonweal in Bolinas, California, and a participant in the Mt. Sinai study, "Many communities, especially highly impacted communities, believe that such information will help them understand their own risks and responsibilities and provide them with information for policy change and regulation" (Patton). For her, body burden testing "is a powerful tool and can galvanize public interest in and support for toxic chemical regulation."

INFANT FEEDING UNDER WRAPS

A look at contemporary culture confirms that most people in the United States do not consider breastmilk or infant feeding significant issues. When we think about the common reaction to breastfeeding, we realize that, for the most part, this event takes place behind closed doors, in "private" spaces; when women do choose to defy the norm and nurse in public, many people avert their eyes out of "respect," embarrassment, or discomfort. Many others make disparaging comments aloud or to themselves, expressing their belief that nursing should be relegated to "appropriate" (private) spaces or that it should end at the "appropriate" time—when a baby is no longer a "baby," has teeth, can walk, or talk.

We tend to construct infant nourishment as a minor part of a host of childrearing decisions, a choice without very significant consequences. Indeed, we regard it as a mother's issue, and a "personal" one that rarely holds the status of other women's issues that feminists have brought to the table including structures of dominance and suppression entailing wide-ranging patterns of social, economic, ideological, and political occurrences—issues such as pay equity, reproductive choices, women's and "minorities'" lack of representation in positions of power, and a multitude of social justice issues. And we believe that infant nourishment concerns only women of childbearing years who choose to bear or rear children, and concerns them only for a narrow window of a few months to a couple of years. Outside of specialized medical, public health, or other journals, attention gets relegated, for the most part, to the pages of parenting magazines and books or breastfeeding advocacy literature. Occasionally, the mainstream press will cover particular aspects, as in the heyday of the Nestlé boycott and the recent PBDE controversy. Those sources that do foreground the subject—from La Leche League's *The Womanly Art of Breastfeeding*, to *The Breastfeeding Book*, to pop magazines such as *Pregnancy*, *Parenting*, and *Mothering*—tend, with the notable exception of *Mothering*, to position breastmilk and infant sustenance as transparent and apolitical, rather than as highly politicized, charged issues with serious import to all of us. They fail to acknowledge that many other issues, including social, cultural, economic, class, and racial disparities, affect this decision.

Some have heard of the Nestlé boycott and the campaign to protect infants in the "developing" world from the formula-producing industry's often unscrupulous marketing practices, but we tend to regard such practices as affecting other people in other countries—not us. In addition, while most of us acknowledge that practices around infant sustenance follow certain trends, historically, socially, culturally, and geographically, most of us do not care to examine the implications of these trends. The reasons for this cloaking of infant nourishment present a complex and engaging narrative. We can locate significant consequences of such cloaking, one of which is that most people still have never heard of the environmental contamination of breastmilk—a serious and telling occurrence both in itself and as a symbol for the impasse at which we, as a culture, have arrived. This book explores the complex reasons for and implications of this situation, and discusses what we must do to adequately come to terms with this state of affairs.

BREASTMILK'S SIGNIFICANCE

So why should we care about infant sustenance? The World Health Organization recommends that all babies should be nursed, preferably for two years, and the American Academy of Pediatrics issued a policy statement in 1997 strongly supporting the physiological benefits of human milk and recommending that all babies should be exclusively breastfed for six months and "for at least 12 months thereafter as long as mutually desired" ("Breastfeeding"). Health benefits include higher immunological functioning, less susceptibility to a range of diseases, and decreased chances of developing allergies, diarrhea, SIDS, respiratory diseases, malnutrition, colic, eczema, ear infections, skin disorders, various types of cancer, obesity, diabetes, Crohn's disease, and a range of other disorders. Breastmilk also positively affects neurological outcomes, immunomodulatory effects, brain growth and development, as well as hearing.

Many studies have suggested cognitive and emotional benefits; for example, some report that breastfeeding renders children more mature, secure, and assertive (Steingraber *Having* 240). While one study suggests that children who nurse have IQ points 8.3 points higher than those fed formula, a more recent study that reanalyzed the results of all previous studies—about twenty—demonstrated connections between intelligence and breastfeeding and found less marked, but still statistically significant differences. Controlling for variables, including parents' education, class, and birth order, it found that breastfed children have IQ scores 3–5 points higher than those fed formula, with the difference persisting when the children reached fifteen years.[5] Other studies have found breastfeeding associated with significant increases in cognitive ability and educational achievement, with results extending well beyond childhood.[6] Benefits for mothers, including adoptive moms, who can breastfeed, include long-term protection from a number of illnesses, such as urinary tract infections, hip fractures, osteoperosis, and breast, cervical, and ovarian cancers. In addition, women gain protection from immediate repeat pregnancies and recover from labor more quickly due to the release of the hormone oxytocin. Public health benefits include a lower risk of hospitalization during childhood and later in life, less parental absenteeism at work (due to fewer childhood illnesses), and less propensity to being overweight, which holds substantial public health benefits given U.S. expenditures on problems remedied by limiting excess weight. Indeed, an estimate conducted in 1995 suggests that if every newborn in the United States were breastfed for just twelve weeks, the healthcare savings would be $2–4 billion annually. Another measure indicated savings of up to $400 per breastfed baby over the first year (Business 8).[7]

One of the most significant aspects of choosing breastmilk over formula has to do with the environmental consequences of formula production and use. Very few of us take into account the environmental expenses of formula production, which includes the drain on resources for the production of cow's milk and soy, and for the packaging, transportation and production of formulas, wastes involved with the dairy industry, use of pesticides for feed, waste associated with the use of dioxins in treating paper used for labels, and other resource burdens. While many of us are beginning to become aware of some of the wider implications of infant feeding choices, we are not encouraged to contemplate these effects on the environment.

Although breastmilk offers many advantages over formula, the didacticism of blaming women who do not nurse involves its own set of problems, including a lack of sensitivity to the fact that many women cannot, for various reasons, opt to use breastmilk. In cases when the mother cannot nurse or express milk, or procure banked milk, formula can literally save lives and offers important health benefits. On the other hand, simply representing this as an equal choice—or, as in the case of the ads featured on *20/20*, representing pro-industry, antibreastfeeding actions as being for the sake of mothers who cannot nurse, comes at a cost to the health of babies, mothers, and the environment at large. This book explores why breastfeeding rates remain comparatively low in the United States, looks at some of the complexities surrounding infant feeding choices, and argues that the best approach would be for us to bring about the conditions so that most mothers could choose breastmilk. And, increasingly, this choice involves environmental pollutants and their effects on breastmilk. Indeed, to talk about infant feeding necessarily involves addressing a range of policy changes that must take place for more women to be able to manage to use breastmilk, and to do so with confidence that their breastmilk will not put their babies at risk.

ENVIRONMENTAL TOXICITY

So what changes would have to take place for all women to nurse with the confidence that their breastmilk is safe? These include enforcing existing legislation limiting the production and use of toxic substances, society-wide recognition that such legislation is woefully inefficient at protecting our environment, our food, our water, our air, and our bodies, and mandating the testing of all chemicals already in use.[8] In addition, we must implement a range of new legislation that would keep dangerous chemicals from being produced and used (including when they are produced as by-products of industry). Perhaps most important, these changes include a commitment on the part of the public at large to do whatever it takes to address environmental degradation.

This book explores patterns of attention and inattention to infant feeding by various groups, and argues for the foregrounding of breastmilk as an important image of the precipice at which we, as a country, now reside. Tracing the trajectory of attention to infant food contamination, I explore the thorny question of why groups that we might expect to embrace the issue—in particular, feminists and environmentalists—have, for the most part and until very recently, remained relatively quiet about the knowledge that environmental toxins have been appearing in human breastmilk. Considering the complexities of such omissions, I address the cross-dialogue between various groups that has brought about a fledgling awareness of and responsiveness to the issue and its implications. Infant feeding cannot be understood or dealt with without a system of analysis that goes beyond a simple reading of the situation. And it cannot be dealt with without a way of conceptualizing that transcends the more oppressive aspects of our current system.

It is important to note that several studies have found that when a woman's body contains higher levels of toxins it is *prenatal* exposure through placental fluid, rather than *postnatal* exposure through breastmilk that is the most damaging (Steingraber; Gross-Loh 63). And some researchers, anxious to curtail anxiety, point out that exposure does not necessarily translate into an effect on the body. While these points are extremely consequential to discussions about breastmilk toxicity, they in no way deny that breastmilk toxicity is a reality. And they do not deny that exposure does occur through breastmilk. While almost all studies indicate that breastmilk, even when contaminated, is by far the best choice for babies, the fact that this pure, living fluid—symbolic of the most essential human connection—is showing signs of environmental pollution should be a wake-up call.

Pointing to the necessity of rethinking how we frame infant feeding, I suggest that as long as we continue to allow breastmilk to be positioned as we have done thus far, we forfeit opportunities to explore ways in which it illuminates our current impasse, which Rachel Carson describes as "a spring without voices." As a substance associated with purity, freshness, innocence, and tenderness, breastmilk occupies a striking position. Although evidence overwhelmingly points to its unmatched benefits, substantial data also suggest that today, in all regions of the world, breastmilk shows signs of environmental contamination. Some studies, for example, indicate that breastfed infants in the United States receive 6,000 times the "tolerable daily intake" of dioxins set by the U.S. government (Harrison 35; US EPA "Health"). Others suggest that each day breastfed infants in industrialized countries, on average, ingest fifty times more PCBs per pound of body weight than adults do—enough to violate FDA standards set for regulated foodsources (Steingrabber *Having* 251). This is significant because of effects associated with PCB contamination, including learning disabilities, behavioral and cognitive problems, neurological and endocrine disruption, and susceptibility to various cancers and other diseases (Colburn 186–97). The Mt. Sinai study showing that the average level of brominated fire retardants in its milk samples was seventy-five times higher than the average for Swedish women and "were at levels associated with toxic effects in several studies using laboratory animals" suggests that it does not have to be this way (Sucher and Walker).[9] That such exposures occur during key moments in the baby's development, when the body's fine-tuned organs continue on their path toward maturity, makes these findings utterly consequential.[10] Due to the process of bioaccumulation (whereby toxins move up the food chain, persisting in the fat of bodies, and magnifying in the bodies of those ingesting entities that have ingested previously magnified toxins), breastfed infants sit at the top of the food chain, ingesting toxins accumulated over a lifetime, including exposures that mothers accumulated from their own mothers. In a very real way, this is an issue that matters not just when we are pregnant or nursing; the toxic load in our bodies begins to accumulate when we breastfeed as babies, and continues when we eat a tuna sandwich at age five, and when we enjoy a swordfish steak along with cake and whipped cream for dessert as adults.

The poignancy of the situation comes alive when we realize that the only way a woman can rid her body of certain toxins built up over a lifetime is to extract them through placental fluid or breastmilk. Again, this issue does not just involve women in their childbearing years; it involves all of us. Although men and non-lactating women do not produce breastmilk, this does not mean that their bodies do not bear the same toxic load as their lactating counterparts. So, in this sense, breastmilk becomes an icon pointing to a bigger picture. It reminds us that even if we, as a world community, got to a point in the future when we choose not to bother with breastmilk and its toxicity and opted for formula on a worldwide scale, believing the benefits of breastmilk to be, in the end, marginal, we would not only have denied future generations the immunological, neurological, and cognitive benefits of breastmilk, we would have merely skirted the issue of environmental pollution. For when breastmilk shows signs of environmental toxicity, so, too, does formula, and so, too, does the rest of the world around us. And when we relegate our infants and children to beginning life with less than the most healthy choice of nourishment, we have allowed a narrowness of vision to diminish their life opportunities, and marked ourselves as citizens who cannot put the needs of future generations above our own shortsighted self-interests.

Before moving on to lay out the basic organization of this book, I want to address one final, important point about breastmilk contaminants. While studies have indicated that all women's milk shows signs of environmental toxicity, many, many studies have documented the fact that certain populations—women of color, particularly Native American women, women living in Arctic environments, poor women, and women exposed to contaminants through occupational exposures—all face higher rates of environmental pollutants than the general population. Their children are the most highly exposed of all people living on earth. Unless we make changes, their children's children will be even more highly exposed. In a very real way, this is also a social and economic justice issue.

LAYING OUT THE PLAYERS

Chapter one, "The Evolving Narrative of Infant Food Contamination: Historical Vignettes," considers how a series of supremacist gestures reside at the heart of contamination narratives. It argues that to tell the story of the "birth" and development of attention to breastmilk and infant food contamination is to register a long history of oppressive, supremacist behavior; it is to witness a series of exclusions, erasures, and silencings. From claims that the milk of the "delicate" and "high-strung" noblewomen in some regions of seventeenth-century England would be less salutary than that of peasant wet nurses, to fears of suckling by "unsavory" immigrant wet nurses suspected of carrying typhoid and marks of a tainted character in turn-of-the-century Boston, to anxiety over HIV transmission in Los Angeles and Nigeria, this story includes representations of infant feeding and approaches to "environmental" health. And what this story suggests is that contamination does not just happen. It follows certain patterns that we have laid out through our collective decision-making. And those patterns illustrate the ways in which oppressive gestures and attitudes lead to oppressive behaviors that determine a series of outcomes, one of which is that the most vulnerable groups—women, people of color, immigrants, people of lower economic standing, diverse species of plants and animals, fetuses and babies—feel the impacts more forcefully than those in positions of power, those who make the decisions that affect all of us, but who, because of their power, have the means to better protect themselves.

Tracing patterns of attention and inattention to child sustenance by various groups suggests that such patterns stem from the ways particular groups get positioned and position themselves in relation to hegemonic structures of power. Here, outsider status, trenchant critiques of dominant discourses, and creative articulation strategies have begun to emerge as effective interventionary tools, thus opening the door to and informing the discursive choices and actions of more "mainstreamed" attention to infant feeding issues. Despite facing a host of complex problems, some activists have begun to fight back.

Chapter 2, "Toxic Disclosure: The Growing Awareness of Infant Food Contamination in the Context of Much-Needed Breastfeeding Advocacy," considers why U.S. environmentalists have not, at least until very recently, chosen to press the dramatic issue of breastmilk contamination, which one might expect to be a "poster-child," flagship issue. Speculations about this failure to act include concern that mothers might refrain from breastfeeding after learning of possible contamination, "personal anxiety about an issue that fundamentally challenges our conceptions of our own bodies and our relationship to our children," and a reluctance to leave themselves open to further characterization as "hysterical fear-mongers." They also include a reluctance to risk promoting formula, out of concern over the environmental impacts of formula production, transportation, and use (Harrison 35).

In the early 1990s, environmental justice activists began to critique mainstream environmentalists for their lack of attention to environmental justice issues, and for racism and classism apparent in the lack of representation by people of color within their ranks. However, recent attention to breastmilk contamination, and to the ways in which indigenous peoples, people of color, and poor people face greater exposures to toxins that lodge in breastmilk indicates that environmental justice critiques of mainstream coverage has begun to pay off. In addition, feminists and breastfeeding advocates have played significant roles in the changing terrain around infant feeding. For example, feminist critiques of science and their insistence that biomedical testing be conducted on women, along with breastfeeding advocates' critiques of formula marketing practices have participated in moves toward greater responsibility.

While most research continues to suggest that breastmilk, even when contaminated, offers far more benefits than formula, if we do not act, there will be a time in the near future when this will no longer be the case. The scales could tip, as they have already done for women with the very highest exposure rates. This issue involves a complex history, including the fact that for many years in the past, in many geographies, dominating discourses and cultural colonization converged to sanction women who choose breastmilk over formula. The low rates of breastfeeding in many areas of the world have contributed to our lack of regard for breastmilk, while our utilitarian approach toward the natural world, seeing diverse ecosystems as "resources" to be extracted, used, and polluted, has contributed to the situation we currently face. This chapter explores those chains of events—and our (in)attention to them—that have brought us to this impasse.

Chapter 3, "Breast Fetishization, Breast Cancer, and Breast Augmentation: The Curious Omissions of Breastfeeding and Breastmilk Contamination as Significant Feminist Issues," turns to the question of why feminists, for the most part, have not embraced infant feeding or breastmilk toxicity as key issues. My research has uncovered complex reasons, including a reluctance on the part of many feminist thinkers to foreground something that invokes the "traditional" nuclear family with its links to fixed gender roles, patriarchal structures of governance and value, and heterosexual privilege. In other words, to embrace infant feeding as a key issue—to rally for workplace accommodations for nursing and pumping breaks or for attention to controversies around formula and breastmilk contamination—is to run the risk of invoking social and economic frameworks whereby the traditional capitalist division of labor relegating mothering and nursing to the domestic sphere have been used to justify patriarchal dominance. Another reason for feminist inattention to infant feeding may be that breastfeeding seems so fully to invoke essentializing frameworks of identity.

Perhaps the most telling case of infant food toxicity involves Native American women living on reservations where environmental waste contaminates the water, the air, the soil, the plants, the fish, and wildlife—and the breastmilk of mothers. Chapter 4, "Polluting the 'Waters' of the Most Vulnerable: Environmental Racism, Environmental Justice, and Breastmilk Contamination," addresses the plight of Native American women and women of color, and how our collective prioritizing has left them, their children, and their children's children vulnerable. Repeatedly and systematically, Native American communities and communities of color have been targeted as the primary sites at which we locate polluting activities; and the stark case of the tainted breastmilk of the St. Regis Mohawk women reminds us that some women's infants bear the cost of inequitable distributions of power more than other women's. But as environmental justice advocates have argued, millions of people throughout the United States (and the world)—particularly people of color and of low-income groups—live in geographies exposed to elevated levels of environmental toxicity. Many studies confirm that polluting industries, transfer stations, toxic waste storage facilities, mining, and other industries all get located more frequently in areas where people of color and lower economic standing live. Their breastmilk is thus subjected to greater risks of environmental toxicity. The situation illustrates how patterns of dominance and oppression work in subtle ways—beginning with exposure through placental fluid and with a baby's first mouthful of its mother's milk. The problems continue with inequitable access to healthcare. Toxic breastmilk marks supremacist and

genocidal practices, reminding us that racism and colonization continue to
operate, affecting people's lived experiences in the most subtle ways. But with
the exceptions of environmental justice activists, breastfeeding advocates,
environmentalists (increasingly), feminists (increasingly), and a few others, we
as a culture do not take notice, and have remained silent.

My hope is that this book will raise people's awareness and help us all
to see that breastmilk matters—not just to nursing mothers, but to all of us.
It reminds us that the toxic load in each human body begins with transmission through placental fluid and breastmilk and continues to build throughout our lives. Just because breastmilk is a site in which bodily toxicity can
easily be measured does not mean that it is more contaminated than the rest
of our bodies. Tainted breastmilk speaks to the widespread contamination of
our entire planet. And it reminds us that such contamination most fully
impacts those who are the most vulnerable. I do not want this book to cause
anxiety over our current situation, bring mothers to abandon breastfeeding,
feed the dismissal of environmental awareness on the grounds that it is
paranoid and overblown, or empower rhetoric that seeks to use critical analyses
of feminisms as weapons to discount the rich contributions of feminists or
undermine continuing feminist work. My hope, rather, is that this book will
inspire more people to think critically, to read events in their everyday lives
with thoughtful attention, and to act on their convictions.

Environmental Defense Attorney Karen Florini notes that while scientists have conducted many, many studies on PCBs, dioxins, DDT and its
breakdown products, and "a handful of other chemicals," most of which have
already been banned, other exposures must be investigated because "This
allows us to shape public policy" (1). I firmly believe that we need policy
shifts. But I also believe that we need to create shifts in consciousness, in
how we engage the world that surrounds us. Stressing that "New
findings . . . would have to be communicated with great care," that "mother's
milk is clearly the best food for babies," Florini argues, that "a national breast
milk monitoring effort should be coupled with aggressive efforts to promote
breast-feeding" (Florini 1). My research suggests that these must be combined with continuing efforts to raise citizen awareness, to sharpen critical
thinking skills, and to bring about greater citizen participation.

Contaminated breastmilk is a wake-up call, jolting us from our complacency, our arrogance in privileging our rights above those of others, and our
narrow-mindedness in refusing to address our impacts on diverse ecosystems,
on biodiversity, and on all generations yet to come. It poignantly reminds us
of a connection: when a species becomes extinct, it is lost forever, and when
toxins impact plant, animal, and human health, those effects extend through
generations. Breastmilk is also an icon of hope and renewal, reminding us
that we can choose to create health and elect to forge justice, not just for
the privileged, but for all.

ONE

THE EVOLVING NARRATIVE OF INFANT FOOD CONTAMINATION

HISTORICAL VIGNETTES

<hr>

Part I: Wet Nursing

CONTAMINATION NARRATIVES: A SERIES OF SUPREMACIST GESTURES

It makes sense to begin to come to terms with the story of breastmilk toxicity by examining a series of historical events. This chapter offers sketches of a few moments in which infant sustenance either occupied public sentiment as being a problem or played a significant role in infant health, yet, for some reason, did not enter public discourse as significant. It is instructive to position the concept of and conditions surrounding "environmentally contaminated" breastmilk within a larger narrative of infant food contamination, and to investigate the degree to which this larger narrative involves the operation of often hidden gestures—what I will call "supremacist gestures"—within broader patterns around infant nourishment. This can help us to assess one of this book's hypotheses—that to foreground the environmental contamination of breastmilk is to confront a series of problems central to contemporary culture—including sexism, racism, classism, anti-immigrant biases, ageism, speciesism, imperialism, and "corporatization"—problems that we must confront if we are to move toward health and justice and away from current trends toward short sightedness, irresponsibility, and injustice.[1] For us to come to a place at which we see breastfeeding—in any situation—as a fully accepted practice, and also one worth protecting, it is important for people to become more aware of the history of breastfeeding, as well as perceptions about infant food contamination. Historical events form the background that has cascaded into a series of situations around infant feeding that we now face.

Part of coming to understand how approaches to infant feeding mask other, more insidious social, economic, and ideological impasses, has to do with coming to understand ways in which many of our actions today mimic, in sometimes subtle ways, practices associated with various historical moments. However the intricate web of events surrounding a particular moment in a particular geography play out, the figure of breastmilk tells a narrative about the structures of domination impacting the real-life experiences of diverse groups of people. I have chosen to offer a few vignettes of specific situations, rather than a flowing or inclusive narrative of the history of infant feeding trends; space does not allow such a genealogy, and that approach, methodologically, might promise something that would encourage the diverse experiences of peoples to be generalized or misrepresented for the sake of offering a cohesive narrative integrity.

DEFINING TERMS

At this point, it makes sense to offer some definitions. Most of the time, when we invoke the terms "environment" and "contamination," we refer to a range of pollutants that we release into the "natural" environment—the air, water, soil, organisms, and bodies that surround us. But we can also read the "environment" in more expansive terms, taking a cue from the environmental justice movement, seeing it to refer to the places where we live—the cities, villages, neighborhoods, reservations, ghettos, and barrios, but also the nations, regions, and state of affairs we inhabit. Expanding the term's meaning brought the beginning of the realization that the effects of our collective choices cause all people's environments to be repositories for dangerous environmental pollutants—but some people's more than others. It has allowed activists, particularly environmental justice activists, to talk about how their bodies have become colonized, marked with disease and ill effects from oppressive policies and practices played out, literally, on people's bodies. Infant feeding occurrences tell these stories, as in the case of the Inuit; these tribal women's breastmilk shows the highest levels of PCBs worldwide, so that mothers cannot live fully in their environments, eating traditional foods and also safely nursing their babies. One's "environment" can also include the webs of social, cultural, and economic events—various structures that come together to shape our experiences, as well as the opportunities we are able to grant our children and our children's children.

And so when I speak of infant nourishment contamination, this might involve pollutants and waste from industrial processes as well as a range of policy decisions governing their release into the spaces where we live. Or it might involve perceptions about infant sustenance; contamination narratives also include moments in which some women's milk becomes tainted, is perceived to be unfit, has some deficiency, or, as in the case of many periods in twentieth-century America, peaking in the late 1950s and early 1960s, when all women's milk began to be seen as somehow inferior to the man-made substitute. One's environment and the degree to which it is polluted, or is seen to be, can involve issues of race and class. People of color, immigrants, and low-income people live and work in spaces that subject them to toxins; in some historical junctures, they have lived under social stigmatization that defines them as somehow less "clean," less morally "fit." One's environment can involve choices about whether one should nurse if one tests HIV positive (in the attempt to limit the transmission of the virus through breastmilk), whether governmental or relief policies or those involving the marketing of formula will affect how one manages the question of possible breastmilk "contamination." It can involve social, cultural, ideological, political, and perceptual questions about gender relations and marketing practices.

Infant nourishment entails structures of oppression and power. It reminds us of how some of us are privileged and some of us are positioned to work toward bringing about a more healthy state of affairs. Infant feeding events remind us that we can choose to begin the long process of starting to honor the rights of all to health and freedom from oppression, or that we can decide to continue on our current path—one in which we allow oppressive structures to affect billions of people's lives. To sort through the question of where the interests of various groups intersect over infant feeding issues, it is important to investigate some of the anxieties, patterns, and habits that informed infant feeding practices at various historical junctures. Indeed, one of my contentions here, is that we need better interface between the various groups that have either shown an interst in infant feeding issues—or that are in the position to impact infant feeding. It is instructive to consider the successes and failures, the discursive battles, and the shifting practices around infant feeding.

We might begin by acknowledging that infant feeding to some constitutes a very particular physical and spiritual relationship that lies at the heart of human societies; it is the nourishment of our infants and children that allows for the perpetuation of the human race, and it constitutes a first gesture of engagement that will become part of a process in which children become socialized, learning to connect with others through language, social engagements, and other means of interaction. However, in saying this, we must recognize that for many disparate cultures at many junctures and in a variety of ways, the connection between infant and caregiver has been something other than a sacred, mythic, mother/child bond in which the mother suckles the child in an expression of deep connection. People have been using "surrogate" feeders and alternatives to breastmilk since the beginnings of recorded history, if not before (Golden; Jelliffe and Jelliffe; Palmer). While one of the goals of this book is to bring the general public to a level of comfort with breastfeeding such that women feel encouraged to nurse in any setting and are confident that their milk is free from contaminants, this does not mean infant care should "return" to some mythical zone in which only the biological mother/

child bond gets sanctioned, or where women are understood to be solely or primarily responsible for childrearing. Indeed, with more and more families opting for adoption, with a range of exceptions to the traditional biological nuclear family taking shape, with ravages of HIV, other diseases, natural disasters, and war causing babies to lose their biological mothers, and with more and more women working outside the home, it is important that other arrangements—surrogate nursing, breastmilk banking, legislation mandating nursuries and pumping and nursing breaks for mothers, fathers, and other caregivers—are instituted and made available.

ADDRESSING THE (COLONIZING) DANGERS OF OVERGENERALIZATIONS

While wet-nursing—a situation in which someone other than the biological mother suckles a child—has been practiced in many different locations and many different junctures in history, it is important to refrain from overgeneralizing about it. Indeed, as several studies have suggested, what we call wet-nursing has been practiced in a range of disparate cultures—from ancient Greco-Roman societies, to some cultures in Europe through several centuries, to contexts in Asia, Africa, and other parts of the globe, to certain enclaves in contemporary social settings (Jelliffe and Jelliffe; Maher). But, as a number of feminist scholars have reminded us, one danger of overgeneralizing about a historical practice is that this method often brings us to gloss over the intricacies of the specific cultural setting in which any given practice has taken place. It is to erase the lived experiences of real people—their struggles, pains, daily habits, and the particularities of their life stories. As Uma Narayan, Chandra Mohanty, and other feminist critics have suggested, failing to take into account the intricacies of particular social, regional, political, ideological, and economic variations is to engage in what Narayan calls a "colonialist representation" (45). Narayan here is critiquing what she calls "ahistorical and apolitical" Western feminist understandings of "Third-World traditions" that "replicate . . . problematic aspects of Western representations of Third-World nations and communities, aspects that have their roots in the history of colonization." She explains that such understanding often involves generalizations made by "Western readers unfamiliar with the historical, social, political, and cultural contexts of the practice being discussed" (46). Her warning involves recognizing that "such contextual unfamiliarity is likely to enable problematic representations to be accepted uncritically and without awareness that the text contains an interrelated cluster of misrepresentations that collaboratively constitute a 'colonialist stance.'"(46).

If we consider wet-nursing as it was practiced in the United States in the seventeenth through the nineteenth centuries, we see that different groups of people situated the practice differently, along a variety of structural axes (Golden 32). Looking at wet-nursing as a social custom, we discover that it occupied different positions at different times and in different places, being at one time an act of friendship, at another a paid labor transaction, and at another becoming part of a system of plantation slavery. Vast disparities informed the practice; it could be an arrangement set up for a brief period or could last for several years, could bring the wet nurse a variety of rates of pay, and could offer her great prestige or be a chore commanded of her, without her ability to refuse. According to Golden, "the continuum of demand ranged from need—arising from the death or illness of a mother—to a choice," while the continuum of employers "stretched from private families to public agencies." Wet-nursing also followed a geographic continuum; some wet nurses lived with their employers, others took babies into their own homes, while still others visited the homes of their employers or the friend whose children they suckled (32). And, when it is practiced today, variations continue to operate.

Gabrielle Palmer, in considering European contexts, notes the "wide differences between regions," pointing out: "In certain areas of Europe artificial feeding was already well established before the nineteenth century," while in others it did not take off until much later. In addition, some women weaned before a year, while in East Lincolnshire women were reported to suckle their children until they were seven or eight years old, even in the 1820s (165). While we want to refrain from seeing all wet-nursing arrangements as being comparable, we can, on the other hand, draw some conclusions based on observances of practices that had similarities in a range of situations over an expanse of time. Palmer, for example, observes that while variations existed "between individual women, households and regions . . . it was certainly accepted for many centuries that important women often did not feed their own babies" (148).

WET-NURSING AND SEMEN-CURDLED MILK: ASSESSING THE MYTH OF THE CONSTITUTION TOO DELICATE TO SUCKLE

Palmer surmises that because some middle-upper-class wives in contemporary U. S. contexts have chosen not to work outside the home "because their husbands could afford it," the habits of middle-upper-class women in many settings around wet-nursing seem to have been, in part, an expression of their socioeconomic status (Palmer 150). Indeed, research suggests that upper-class women viewed nursing in the same light as hoeing the fields or working the spinning wheel—as a task they would never think of performing. While such a parallel may allow us to understand historical situations through the filter of more familiar practices, wet-nursing in many settings appears to involve more than the simple fact of social rank. The more interesting and telling aspects of this story have to do with the specifics surrounding the long list of reasons why these women did not tend to nurse. It is not just that these women were considered too privileged to engage in suckling their own children, but that they were assumed, in many cases, to be too delicate, high-strung, nervous, and excitable to produce milk that was either acceptable or adequate. Palmer notes, "One reason for discouraging noble-women from feeding was that they supposedly lacked the desirable placid temperament that would be passed on to the baby through the milk" (161).

Feminist scholars Ehrenreich and English, writing about some settings in eighteenth- and nineteenth-century England, suggest the degree to which certain cultural codes informed thinking about nursing for a long period of time and in a range of settings. They note: "It was as if there were two very different species of females. Affluent women were seen as inherently sick, too weak and delicate for anything but the mildest pastimes, while working-class women were believed to be inherently healthy and robust" (Ehrenreich and English *Complaints*, quoted in Palmer 161). A comment from 1656 London on adult suckling (which was employed in certain situations of adult illness) illustrates the degree to which the health and purity of breastmilk were seen—at least in some settings—as linked to the diet, but especially to the disposition of the suckler. J. Beadle comments in his diary that "what made Dr. Cajus in his last sickness so peevish and so full of frets at Cambridge, when he suckt one woman (who I spare to name) forward of conditions and of bad diet; and contrariwise as quiet and well, when he suckt another of contrary disposition; verily the diversity of their milks and conditions, which being contrary one to the other, wrought also in him that sucked them contrary effects" (109). Some women, quite frankly, were seen to produce tainted milk.

Several scholars have noted that physical conditions certainly played a role in perceptions about the relative merits of various women's milk and perhaps rendered the breastmilk of some noble and upper-class women less abundant or salutary. For example, for many generations, peasant women in many social settings, because of gaming laws and the economics of food distribution did not have access to rich meats, liquors, and a variety of refined foods accessible to wealthier women, so that their diet of vegetables, whole grains, and legumes, depending on the region, would have offered what some would have considered to be more healthy lactation fare. According to Palmer, "The rich woman lying on her couch in the drawing room and perhaps only venturing forth in a carriage, lest the sun freckle her lily-white complexion, might have been vitamin D-deficient" (161). In addition, many peasant or lower-class women in many of these settings could not have afforded a doctor so that they would not have been subjected to the medical intervention of bleedings, which health professionals used liberally—"par-ticularly for pregnancy and post partum conditions including perinatal haemorrhage" (161). This treatment, in many cases, would have exacerbated the chronic anemia that many non-lactating upper-class women faced as they proceeded through miscarriages, pregnancies, menstruation, and more pregnancies. That infant mortality rates remained lower in the agricultural regions than elsewhere up to the twentieth century suggests that the necessity for working women to spend much time outdoors would have protected them from vitamin-D deficiency and other health issues that might have affected their children (63).

Aside from possible physical differences, the operation of cultural codes certainly impacted practices, but in complex ways. Palmer notes that while cultural rules meant that the noblewoman was "expected to delegate all physical labor to others. . . . Eventually the myth that noblewomen were too delicate and special to suckle would have provided strong emotional inhibition; how could she know that she could feed if everyone presumed she could not?"(148). But there is more to the story. When we look into the discursive constructions of infant feeding prevalent at several junctures and in many regions, we see that behind this fear of the milk produced by these "delicate" women lay another fear—that the semen produced by these women's husbands curdled their breastmilk, so that, in many cultural settings, one finds a taboo against a woman engaging in sexual relations while she is nursing (Palmer 152; Pollock 215).[2] This fear of semen-contaminated breastmilk, which mandated abstinence during lactation, provided a powerful incentive for women of the upper ranks of the social scale to refrain from nursing; indeed, it may have played a key role in sustaining the practice of sending one's children off to be nursed by another. Linda Pollock's research suggests that the main reason for wet-nursing was the prohibition against sexual relations and the "belief that semen was supposed to curdle breastmilk" (215). But behind this, one discovers a perhaps more powerful group of incentives for the employ of wet nurses—the likelihood that women knew that nursing limited fertility, and the pressure for women in higher social ranks to produce heirs. This reading is corroborated by the apparent disregard concerning the degree of abstinence from sexual relations among the peasant (wet-nursing) classes, suggesting that semen-tainted milk was not the primary disincentive.[3]

In other words, it seems probable that the real fear on the part of an individual man of high social rank may not have been that his wife's breastmilk would be tainted by semen or by her "weak" and "delicate" constitution— however much these fears tended to get circulated in the public domain— but rather that if his wife suckled their child, she would not be able to produce the abundant offspring necessary to ensure carrying on the family bloodline. Indeed, much of the literature of the day points to the intense pressure to produce heirs, particularly in light of the high infant mortality rate for babies of the wealthy—a rate we can now link, in part, to the lack of adequate spacing between births, which would have been exacerbated by the failure to suckle and thus suppress ovulation (Prior 27). Some of these women certainly realized that nursing would reduce their fertility and increase the likelihood for more of their infants to live; but the literature then repeatedly suggested that the pressure on these women to bear more and more children seems to have outweighed any desire many of them might have had to breastfeed for this purpose (Prior 27). Palmer corrects what she calls a "falsehood of history—the myth of the poor 'breeding like rabbits,'" and claims, "on the contrary, it was the aristocrats that deserved this comparison" (151). Indeed, in pre-industrial England, it was not uncommon for a noblewoman to endure eighteen pregnancies. Ann Hattton, for example, a not so atypical wealthy seventeenth-century English heiress, had thirty children: five sons, eight daughters, ten who died young, and seven who were stillborn (151). Corroborating this picture, when wet-nursing had to do as much with the primogeniture laws and the need to produce heirs as with the actual fear of milk contaminated by semen or made unsuitable by the noblewoman's weak, nervous constitution, we find that "as the Renaissance advanced, the image of the nursing Virgin waned in popularity" (Dryden 166). The entire social fabric around infant feeding shifted in response to structures around economic and inheritance practices.

CLASSISM, SEXISM, AGEISM, AND LOOKISM: ASSESSING SOME TRENDS

While the dominant discourse around infant feeding for many centuries in many settings in Europe and the United States revolved around the understanding that women of rank could not produce the hearty, safe milk offered by their more robust sisters, reading between the lines in less dominant discursive zones such as journals, commonplace books, and advertisements suggests that yet another concern lay behind this class-based distinction: the desire among landed women to maintain a certain socially prescribed shape. At some junctures we witness anxiety among noble women that nursing would "ruin" their bodies, and make their breasts less beautiful. Another concern that emerges is fashion. By the sixteenth century, women of means began to wear corsets of leather, bone, or metal that flattened the breasts and nipples, making nursing virtually impossible (Palmer 150). On the other hand, poor women remained—at least until the Industrial Revolution reached full tilt—relatively unaffected by these trends. According to Palmer, "being fashionably dressed was confined to an elite who typically had to do extraordinary things to emphasize their eliteness"(150). Indeed, when one of the rare voices in favor of the English aristocracy's refusing wet-nursing and suckling their own appeared in 1622—Elizabeth Knyvet's book *The Countess of Lincoln's Nursury*—the author made her plea by attempting to "sweep away all possible objections that nursing [is] troublesome . . . noisome to one's clothes, makes one look old, and endangers health" (79).

Women in many of these situations found themselves subjected to "men's laws." And men's laws, bound by structures involving class, ownership of land, title, and blood relations, determined that the pressures on the upper ranks of women to produce heirs and to look a certain way became so great that social practices associated with these goals outweighed other concerns— including protecting women's health, infant's health, and the possibility of mother–infant connections. Even if we dismantle notions of any kind of essential connection between mother and child, and grant, as some contemporary discussions of motherhood have done, that some mothers simply do not feel a "natural" connection with their children, it still seems useful to note that at least some women who were prevented from suckling their children, or even from having their children in their presence, must have felt the loss (in many cases, acutely). Sexism and classism have functioned in many situations in such a way that women have been prevented from maintaining connections with their children. For instance, one has only to think about the practice in some contemporary settings of authorities depriving women of their children because of strict welfare policies. Some early accounts suggest that women suffered on an emotional level from separation from their children, while many others imply that, at times, the strain of frequent and excessive pregnancies in the name of producing an heir meant that many women and their babies died from childbirth and other complications. Without a doubt, infant feeding practices in many settings had significant bearing on maternal and infant mortality rates, and limited some women's lives.

Accounts of effects on wet nurses vary, with studies suggesting that many of them certainly found themselves climbing social ranks through association with their charges. After all, in some periods, wet-nursing became one of the few professions available to women, and one in which wet nurses could gain significant social standing and economic independence.[4] It should also be noted that wet nurses often worked for many, many years; Judith Waterford offers a powerful example of the way such work served to bring some women into social settings that would otherwise have been closed to them. She worked as a wet nurse for over forty years, suckling a number of babies of prominent families. On the other hand, the literature suggests that some wet nurses of various periods faced problems. In some cases, these women became subjected to disease transmission through the process of suckling sick babies. In addition, some accounts have hinted that at least some of the wet nurses' own babies died for lack of proper nutrition as a result of their mothers being employed to feed babies of the well-to-do. Janet Golden points to the classism exposed by an 1894 novel *Esther Waters*, in which the heroine articulates "what physicians had long admitted—that wet nursing often involved trading the life of a poor baby for that of a rich one" since wet-nursing often exposed the nurses' own infants (97). Other accounts suggest that wet nurses often sent their own children out to be suckled by wet nurses lower on the social scale, and that while this practice endangered some, the implication that wet nurses were risking the lives of their own children is sometimes overblown in the literature.

When Mary Wollstonecraft forged ahead with her crusade to end the "subjugation of women," she based her argument on the exposure of the class-based assumption that women of wealth were too delicate to do anything except pursue the most mundane and insignificant pastimes, and she rallied against such socialization by arguing that women should come to terms with real-life responsibilities and become full citizens. Wollstonecraft takes up breastfeeding as a central tenet of her argument: "The wife, in the present state of things, who is faithful to her husband, and neither suckles nor educates her children, scarcely deserves the name of a wife, and has no right to that of citizen." But she does not blame only women—she faults the system of socialization, the "respect paid to property," from which flows "most of the evils and vices . . . as from a poisoned fountain." And she blames men, proclaiming: "Would men but generously snap our chains, and be content with rational fellowship instead of slavish obedience, they would find more observant daughters, more affectionate sisters, more faithful wives, more reasonable mothers—in a word, better citizens." Speaking as a dissident, and from what would remain the margins of social understanding for a long while, Wollstonecraft looks forward to a time when babies would no longer be "sent to nestle in a strange bosom." It is interesting that Wollstonecraft, in exposing the class-based nature of practices that kept certain women from nursing their own children, should use an image of a "poisoned fountain" to articulate problems springing from the injustices of class and gender-based oppression. The labels of "poisoned," "polluted," "curdled," or "unfit" breastmilk emerge in discourses that for a long time in many regions in Europe and America kept many upper-class women from suckling their own. Wollstonecraft suggests, through her image, that in many of these situations, while certain breastmilk was characterized as "poisoned" or "contaminated," it was, in fact, social relations or class politics that were "poisoned," so that, in some cases, wealthy women's breastmilk got abandoned as "unfit," while in others, as we will see in the next section, poor and immigrant women's breastmilk became scapegoated as "polluted" or carrying "blights."

THE "THREAT OF THE DANGEROUS STRANGER": WET-NURSING IN SELECT MID-NINETEENTH-CENTURY AMERICAN CONTEXTS

As in earlier periods in various regions, the belief, in the middle of the nineteenth century and soon after in the United States, that middle- and upper-class women's weak constitutions and "placid temperaments" meant that they could not produce milk fit for babies distinguished popular and medical literature (Golden 151; Palmer 149). For example, Ticknor, in *A Guide for Mothers and Nurses* (1839), argues that "women of the higher classes frequently possess such extremely sensitive and excitable temperaments as will render it imprudent for them to suckle their own children" (92, quoted in Golden 53). Indeed, a study of 1,000 women in Boston in 1908, and published in the *Journal of the American Medical Association* that same year, reports that 90 percent of 500 women described as poor and living in tenements were able to nurse their babies for nine months, whereas only 17 percent of 500 prosperous women living in better parts of the city could do so (Snyder 1213, noted in Golden 139). Clearly, while the availability of other means of feeding their children had an impact on wealthier mothers' "inability" to nurse, the fact remains that the widespread discourses telling them they could not produce healthy or sufficient milk and the women's beliefs about themselves played a role in their (in)ability to nurse.

The practice of the wealthy employing wet nurses persisted in many areas—but with a difference from earlier times. Because of demographic and social changes brought on by the Industrial Revolution, increasing urbanization, and shifting class structures, mothers stopped sending their babies to the country and started having wet nurses live and work within the home (Golden 38; Palmer 168). The practice of locating wet nurses as live-in domestic servants meant that a new population of employees came into the wet-nursing business. Wet-nursing had previously been performed by rural married women with homes of their own, but it now became a temporary occupation for poor, urban mothers. And as a new class of urban poor replaced the peasant matron as the typical wet nurse, the gulf between employer and employee grew larger and more troubled, with the new wet nurses scapegoated as responsible for infecting the babies of the wealthy with a host of blights. These women's breastmilk became associated with disease, including scrofula, tuberculosis, typhoid, syphilis, and "wasting diseases." And the public began to fear that the perceived lax morals, intemperance, and unmanageable "ill nature" of these women would spread through their breastmilk. Increasingly, the public began to express suspicion of this new class of wet nurses. As the new wet nurses became more of a fixture in their employer's homes, they were more and more associated with the taints of unwholesome environments brought from the slums and tenements, and with potential threats of "flawed" characters being imparted to babies (Fildes 168–210; Golden 38).

In England and the United States, the new wet nurse was more likely to be a younger, unmarried mother; the lack of trust was reflected in the fact that "this wet nurse had to live in and be completely separated from her offspring in case she was tempted to feed it" (168). With the problem of the displacement of large segments of people from rural areas, particularly after the enclosure acts privatized the commons that had provided many with opportunities to engage in subsistence activities, many women, cut off from rural communities and thrust into urban slums, found themselves with few choices outside of prostitution and wet-nursing. Prevailing notions were that "fallen women proved by their very condition, that they possessed uncontrollable, ill-disposed emotions, which affected the quality of their milk and was even supposed to convey cancer" (Palmer 169; F. B. Smith 83). Public outcry became heated as people worried in newspapers, magazines, and pamphlets that such women's "tainted" moral characters might infiltrate their babies.

Similarly, in the late nineteenth century, many antebellum physicians believed that strong emotions made milk "toxic" (Golden 151). Legends circulated among professionals telling of mothers and wet nurses who had "witnessed acts of violence, engaged in sexual intercourse, or otherwise become excited." When babies began to nurse from "the breasts of those subjected to such vagaries of poor environments," they "fell into convulsions and died shortly thereafter." A popular novel, *First Baby*, published in 1881, suggests the degree of anxiety over the problem of milk tainted by the character of the wet nurse: when the mother in the book suffers pneumonia, the father and his doctor get names for wet nurses from the *New York Herald* (Golden 141). After searching and searching, they finally hire Mrs. Badall, who is promptly fired after the employers fear that a visit from her drunken husband had rendered her milk unfit. The novel points to the view of human milk as a "volatile substance that could become toxic as a result of the lactating woman's strong emotions and, by implication, sexual activity" (Eberle 35; Golden 66).

While these "dangers" to infants elicited what at times became a heated public anxiety, another danger—the transmission of syphilis—was sometimes suppressed so that wet nurses occasionally contracted the disease from their more privileged nurslings, even as they were frequently scapegoated in various media as being the only source of the disease. As at least one study suggests, the assumption that the "moral taint" could be spread only from the lower to the upper ranks of the social scale resulted in little being done to protect the health of wet nurses, many of whom were thrust into wet-nursing by strict welfare policies that gave them few other options for making a livelihood (Golden 81; 144). Since estimates of infection among the "better classes" ranged from 6 percent to 18 percent in the early twentieth century, wet nurses "had reason to worry"; but this part of the story, for the most part, remained silenced (Golden 145; Jeans 55). The policy of restricting welfare eligibility to force individuals to work, and the belief on the part of civic officials that if a woman "had milk to sell" she should not receive a stipend (called outdoor relief), meant that women had little choice in the matter (Golden 81). An exception can be found in a play called *Damaged Goods* about an upper-class baby who is infected with syphilis; the climax occurs when a doctor "saves"—at the last minute—the lower-class woman who had been hired as the wet nurse (147). What is telling is that although bottle-fed babies—who became more prominent at this time—could and did die quickly of infections caused by bacteria-laden milk, that which Golden calls the "slow and elusive venereal infections were even more dreaded, not least because of the social stigma attached to them" (143). When people denounced wet nurses, descriptions involved "the rhetoric of moral pollution" (155). Immigrant women serving as wet nurses were blamed for typhoid fever, and shunned for fear that they would translate their fallen and tainted morals to infants. As with syphilis, "rates of infection" suggest fears had more to do with racial, social, and economic factors than with an easily stratified threat from wet nurses.

One reason for the shift in attitudes toward wet nurses seems to come from shifting attitudes toward immigrants, who increasingly served as wet nurses. First Irish and later Scandinavian and German immigrants replaced native-born servants (Golden 43). Occurring at a time when "nativist" concerns pervaded public sentiments, this shift determined that such women, considered racially and ethnically "other," were subjected to social stigmatization and scapegoating (Fildes 188–209; Golden 43). To contextualize such occurrences, it is important to bear in mind that "racism" concerns the social stigmatization and oppression of particular ethnic groups based on particular social definitions of difference; as a social construct, it has less to do with actual racial differences than with social constructions of difference at a particular juncture. During the latter part of the nineteenth and early twentieth centuries, nativist sentiments, the privileging of "native" Americans (which meant not Native Americans as we use the term today, but the Anglo-Europeans who had resided in the United States since the early days of colonization), defined many people's conceptions of difference. As the nativist movement took hold, anti-immigrant sentiment became heated. Initially treated as nonwhite, European and other immigrants worked to "deflect" "debate from nativity, a hopeless issue" for them, and to focus on "race, an ambiguous one" (Roediger 189, quoted in Rosenblum 15). So, during this period, anti-immigrant and racist attitudes merged to scapegoat all those constructed as "other." And all of this became tied up in debates over "tainted" breastmilk.

A glimpse of a letter written in 1861 by Elizabeth Cabot, a wealthy Boston matron, suggests how the "geographical and social distance between the working poor and the middle and upper classes was expanding" (Golden 43). She notes: "I roused up and trotted over, and thought I would raise a wet nurse in the village . . . invaded four Irish mansions, succeeded in raising a wet nurse and a woman to take her baby and sent her off with Powel into the town to be examined by Sam [the doctor]." But as women looked to immigrants to fulfill wet-nursing duties, the gulf widened. Golden points out that the literature "refracted the theme of increasing class estrangement by consistently juxtaposing the nurturing middle-class mother against the threatening lower-class wet nurse" (47–48). Indeed, we find that a new concern overshadows the pages of books written in the 1840s and 1850s: the threat of the "dangerous stranger" serving as a wet nurse (Beach 631; Golden 51).

Golden notes how the subject "inspired both the popular and the medical imagination, combining the growing alienation of the middle and upper classes from the urban poor with more specific fears about disease" (Golden 51). We find, for example, that the 1848 edition of the *Home Medical Guide* written by the botanic practitioner Wooster Beach uses the term "stranger" in reference to wet nurses who communicated "loathsome and fatal *diseases* and gave milk 'rendered unwholesome by age or other causes'" (631; my italics). Racism surrounding the arrangements following the Civil War also impacted the scene, so that fears of freed slaves combined with fears of immigrants and created an alarmist climate leading up to the eugenics movement of later decades. It is not surprising, therefore, that in the years after the Civil War, "Nativists . . . complained that Americans of 'good stock' were giving birth to fewer children than in past generations and that those few infants were being nursed at the breasts of immigrants" (137). That lactation suppresses ovulation meant that those engaged in public commentary around "nativism" would not have wanted American women of "good stock" to nurse, since it would have further affected what was seen as "race suicide." That many remained suspicious of immigrant wet nurses most probably served as an underlying motivation for the eventual switch to formula. Many critics "expressed deep concern that foreigners, including foreign-born wet nurses, were corrupting society" so that wet-nursing became a site at which many cultural fears became played out (Golden 153; Haller 53).

Research corroborates that this scapegoating of wet nurses seems to have been a long-standing practice, suggesting that classism, racism, and ethnocentrism have long influenced infant feeding decisions and thinking about the effects of infant nourishment (Fildes 188–210; Palmer 168–70). Indeed, Palmer's comment on earlier arrangements in Europe holds true in the context of many mid-late nineteenth-century situations in the United States: "In spite of the obsession with lineage and 'blood', these noble biological parents apparently influenced only the positive attributes of the child and any adverse characteristics could be blamed on the wet nurse" (161). The classism, sexism, and anti-immigrant sentiments operating here will emerge in other forms in later infant feeding events. While some transmission, or at least perceived transmission, of a range of conditions—from typhoid, to tuberculosis to syphilis—"through" breastfeeding certainly did occur,[5] what is striking is the degree to which contamination discourses involving the "strange others" infiltrated the public domain. Indeed, the degree of anxiety seems to far outweigh the actual danger, especially when we take into account that other, more significant dangers of infant feeding barely registered in public perceptions, if they registered at all. One of these was "swill milk"—post-distillery slop used as the basis for infant formula, a practice protected by governing bodies with an interest in supporting industry, even at the cost of endangering infant health.

The prevailing fears about tainted milk seem to have been linked to a host of social concerns, including shifting demographics, changes in relationships between the poor and the wealthy, unstable social and economic relations, instability resulting from continued colonization of vast portions of the globe, and labor unrest. Several scholars suggest that the proportion of incidences of children dying or becoming ill from unsafe breastmilk did not rise in this period. In fact, more children were surviving infancy and toddlerhood than in previous centuries. But public discourse tended to display a growing distrust of the typical wet nurse, and her breastmilk, thus suggesting that the discourse itself serves as a gauge of other factors that may have had little to do with infant nourishment.

Another indicator of the way class issues get reflected in infant feeding events involves the effects of wet-nursing arrangements on the children of the wet nurses themselves. When wet nurses started to be brought into the homes of the families whose children they were hired to nurse, the danger of their own children being abandoned, becoming ill, or dying seems to have become more pronounced. The public anxiety over the well-being of more well-to-do infants dominated the discourse, while, behind the scenes, the more pronounced problem of the ill effects on the children of the wet nurses received little attention. This suggests, along with the seemingly overblown attention to the possibility of privileged babies contracting strange diseases, that class issues and pressures over unstable social, demographic, and other factors contributed to the anxiety that got manifested as fears of contaminated breastmilk.

This blaming of the less priviliged, and attributing any negative effects to their biology or cultural traits, are relevant as we move to more contemporary contexts. For example, although studies have suggested that the breastfed children of certain racial, ethnic, and underprivileged groups have been more susceptible to negative health outcomes due to their proximity to polluting industries, public sentiment has appeared more concerned with health effects on the breastmilk of more privileged moms—if we can take coverage of the issue in major newspapers and other media sources as a gauge. At other times, classism and racism have emerged in concern over the tendency of poor women and women of color to nurse at lower rates than other women—without attention to the effects of structural racism, classism, and sexism on infant feeding. An article covering a 1995 midcourse report on the status of healthy people 2000's objective, with a follow-up report conducted in May 1999, notes that while breastfeeding rates rose for many women, for American Indian and Alaskan native mothers, the rate decreased 2 percent from 1990 to 1997. Celebrating that other women were nursing at higher rates in 1997 than they were in 1990, the article does not mention that the period at stake coincides with environmental justice groups' attention to health effects of environmental pollution, especially for Native American and Alaskan women. In addition, while the media focuses on the lower breastfeeding rates for poorer women and women of color, they rarely note that these women are less likely to be able to breastfeed, given that policies protecting workplace pumping tend to favor more privileged women. In other words, the public dismisses such failure to nurse as stemming not from an awareness of environmental effects or social effects closely linked to the classism, racism, and ethnocentrism that allowed these groups to be situated in harm's way, but from some flaw—social, cultural, or biological—attributed to the mother herself.

As racial, ethnic, and class conflict and wariness of immigrant women grew, a parallel development—and the degree to which the relationship was causal continues to be unclear—was the increasing intervention of medical practitioners into the practice of wet-nursing (Fildes 168–88; Golden 128–56). As several scholars have noted, physicians such as Dr. Beach focused more and more on problems with infant feeding, linked them to an expanding perception that the well-to-do had become vulnerable to the infections of lower classes and racialized others, and implied that the solution lay in the studied application of medical knowledge (Golden 52). But again, classism, racism, and ethnic prejudice infiltrated this domain and determined the contours of the proscriptions doctors delivered to women. For example, when poor urban wet nurses began to request access to the wealthy woman's diet, Golden finds that "doctors could barely disguise their contempt for poor women who wanted to dine like their social superiors, although they wrapped their analyses in medical terms" (148). John Price Crozer Griffith argued, for example, that "a woman from the lower walks of life given unrestrained opportunity to indulge freely in food to which she has not been accustomed" was "liable to indigestion" and would be more likely to cause problems for her nurslings (Griffith 186 quoted in Golden 148).

Doctors and commerce clearly did play a role in the waning of wet-nursing as a means of employment for women. The gendered component comes into relief when we consider this aspect of the story. In many situations before the latter part of the nineteenth century, "the ability to suckle an infant lifted women out of severely regimented institutional life at almshouses and welfare institutions—once they had become fallen—into relatively high paying positions" (Golden 79). Whether the change can be attributed to perceptions about "polluted breastmilk," to the increasing medicalization of wet-nursing that slowly gave way to the medicalization of infant feeding and the touting of "artificial food," or to the increasing infiltration of "artificial food" into the markets, wet-nursing began to decline.

HUMAN MILK BANKING

Increasingly, the practice of selling milk gave way to the practice of giving milk. And this may reflect a gender bias as well. Certainly, class issues play a role. It is telling that as breastmilk went from being seen as a "commodity" to being viewed as a "gift," the public anxiety over tainted breastmilk began to fade. One marker of the shift occurred during World War II; according to Golden, "the propriety of women selling something that babies needed for their survival began to be challenged during WWII, when blood donation started to be touted as a patriotic duty" (203). An interesting counterpoint comes with the story of Charles Drew, the doctor who developed the technology for blood transfusions that saved many lives during the war and since. The catch is that although Drew was a person of color, the blood of soldiers of color and whites was kept segregated during the war and in later years (Love). So while breastmilk then came to be seen as part of a moral trans- action, something not to be subjected to the harshness of commodification, blood, at the same time, while regarded as part of a moral transaction, some- thing that one could give as a gift to those in need, also became subjected to the calculations of racial prejudice and bigotry.

As milk selling decreased as a practice, milk giving, and indeed the remnants of milk selling when it does take place, seemed to become free of the fears of contamination that had dominated wet-nursing discourses in the latter part of the nineteenth century. Indeed, for a period in the United States in the 1960s and 1970s, graduate students became the predominant sellers of breastmilk; according to Golden, the perception was that these women were selling "smart milk" (204). Perhaps with the foregrounding of class, race, and ethnicity-based anxiety during the civil rights era, the wan- ing of anxiety over the possible contamination of certain women's breastmilk occurred. Milk banking, which continues today, commonly gets framed as an activity free of fears about class and race; the transformation of these issues associated with earlier periods of the movement to subtler, less overt forms of class and race bias—reflected in the dubbing of "smart milk"—seem to suggest a lessening of public fears of contamination. But such apparent con- fidence appears to have masked other fears—for example, the fear of the spread of HIV/AIDS through breastmilk.

The growing recognition of the possible transmission of HIV/AIDS through breastmilk has had dramatic bearing on human milk banks. The Human Milk Banking Association of North America notes that in the early twentieth century in the United States, milk banking "blossomed and grew with increased use of donor milk for ill and premature infants" ("Human Milk"). Early on, mothers with abundant milk were asked to provide suste- nance for premature and ill infants by either nursing directly or expressing milk. In 1909, Austria established the first milk bank, in Vienna, with Germany following soon after, and the United States establishing a bank in Boston, in 1919 ("Human Milk"). As obstetrical and pediatric procedures advanced, and more premature babies survived, donor milk banking pro- gressed, so that, by 1982, an estimated thirty milk banks operated in the United States. In 1985, a group of health care providers and concerned citizens established the Human Milk Banking Association of North America (HMBANA), focusing on promoting milk banking and establishing stan- dards for all North American milk banks. Published in 1990, these standards inform milk banking documents worldwide, with HMBANA representatives reviewing and updating them annually.

As with earlier contamination dramas, the dawning of the HIV/AIDS scare had a marked and immediate effect on public perceptions, profoundly influencing human milk banking. According to HMBANA, "Concern for the unknown and need for increasingly complex screening of donors and milk processing resulted in many banks closing, almost overnight" ("Human Milk"). By the end of the 1980s, the number of banks in North America had fallen from about thirty to about eight.[6] But another shift occurred in the 1990s with evidence of safety and increased research on the benefits of human milk. Many more health professionals and families, particularly those with ill or premature infants, have requested donor milk, and most countries in the "developed" world have established banks. Today, banks pasteurize milk with the Holder Method and check for bacterial growth; contaminated milk pools, which contain milk from three to five donors, are discarded ("Human Milk"). Workers freeze the milk and ship it overnight to hospitals and individual recipients at home, so that many babies who would not otherwise benefit from the unique properties of human milk are able to do so. My hope is that milk banking will become much more popular and that donating and selling milk will serve as ways for women to participate in fostering health and a sense of communalism.

WHITE SUPREMACY AND THE COLONIZATION
OF BLACK WOMEN'S BODIES

Breastfeeding practices that occurred at various moments during the period of the diaspora from Africa reflect how notions about race, power, capital, and labor influence people's lives in myriad ways. During slavery, when enslaved people's bodies become colonized, the slaves' bodies became marked with the effects of violence and oppression through a system of capitalist production, so that they become treated as machines, objects, or animals— as nonhuman (legally three-fifths a human-being). And, while the image of the bound and shackled slave, the bleeding slave, the slave forcefully raped or starved by the slave master haunts us just as later images of lynchings do, other events played out on the bodies of slaves tell other stories of oppression. While infant feeding practices do not disturb us with their violence, they do reflect how the dehumanization endured by slave populations took many forms.

One of the most horrific realities of the system of slavery as it operated in the United States was the frequent separation of slave women from their babies. This served as part of the effort to mark slaves as capital, as "nonhuman" and thus not worthy of being allowed to maintain family units, so that slave children, as capital, could be sold to other slave masters. This separation—and the separation of women from their babies as a result of infanticide or infant mortality from harsh living conditions—stands as a reminder of just how inhuman, chillingly heartless, and calculated such a system was. Before I had a baby of my own, I do not think I could have understood just how unthinkable such a separation could be. Now, reading slave narratives and confronting the systemic separation of children from their mothers, I find these stories even more poignant. In a very real sense, the massive colonization that took place during the slave trade involved the colonization of people's bodies in countless ways, and one that we rarely confront because patently violent images are more compelling includes this institutionalized separation of women from their children. While research suggests that the colonized Africans nursed their own babies before the diaspora, the disjucture of slavery disrupted this practice and brought, in many cases, absolute separation between mother and child.

Biracial images of white women suckling slave babies—to protect their "investment"—and of slave mammies suckling the white masters' babies as one of their duties—babies who would grow to become masters themselves—appear from time to time, reminding us of the absence of the image of mothers—of white or African descent—suckling their own (Fildes 128–29; Golden 26).[7] Also, in referencing biraciality, these images gesture toward the frequent violation of slave women's bodies by slave masters, so that the absent image of the slave nursing her own child becomes fraught with reminders of the constant violence being played out on her body and the lack of autonomy she experienced. Henry Giroux's commentary on the United Colors of Benneton's use of racially charged photographs to sell clothing reminds us of just how provocative and enduring such images can be (23–29). The 1980s era advertisement with a very black woman suckling a very white baby drew attention to the ways in which the company sought to capitalize on charged racial issues, in order to promote their mission—of selling sweaters. Whether championing themselves as somehow uniting various races into a harmonious whole or as portraying some of the most sensitive racial occurrences in contemporary society, with roots extending back to the period of slavery, the image of the biracial suckling registers more than just a welcome reminder that even nursing can become a way to "unite" the diverse "colors" of the globe. While some white women in the upper ranks of American society during many historical and temporal geographies undoubtedly found their infant feeding choices limited, whether they experienced wet-nursing arrangements as deprivation is not unambiguously evident. In contrast, the curtailment among the enslaved could only constitute one of myriad examples of the deprivation and dehumanization played out upon the completely colonized bodies of the enslaved.

Part 2: Mathematical Formulas

THE ADVENT OF SCIENTIFIC INFANT FOOD AND THE ERA OF "SCIENTIFIC MOTHERHOOD"

While most infants in the United States received breastmilk, either from their biological mother or from a wet nurse through the nineteenth century, increasingly—though unevenly—artificial feeding went through a transition from being viewed as a "death warrant" to being seen, by the 1940s, as the saving grace that set babies free from a host of diseases and ailments (Apple 4; Baumslag and Michels 127; Palmer 203). Reviewing this transition from breastmilk to formula during the late nineteenth and early twentieth centuries, and following it to the point when fewer numbers of women (in the United States) nursed than ever before or since—the late 1950s and early 1960s—when hardly any still nursed at six months—is important in that it sets the stage for today's situation. We can link the sluggish dawning of awareness that the environmental contamination of breastmilk is something to be taken seriously to the relatively low rates of breastfeeding.

In part, the shift from breastmilk to formula can be attributed to a shift in attitudes toward science, toward motherhood, and toward the burgeoning medical profession. It is important to note that we can draw a connection between pro-science, pro-technology, pro-capitalistic attitudes—approaches that often privilege these without due attention to systemic consequences—and the assumptions that have brought us to the current impasse in which the environmental contamination of our water, air, soil, bodies, and breastmilk is a reality (Carson; Merchant *Death*). It is worth reiterating that for people to become concerned about the environmental contamination of breastmilk, they have to first see breastfeeding as an important activity. And, as this section will suggest, a chain of events in the later half of the nineteenth and early part of the twentieth centuries determined that the vast majority of U.S. women by the early 1960s would never nurse their babies.

During this period, a developing confidence in science, medicine, and technology, discoveries in bacteriology, anatomy, physiology, and nutrition, as well as changes in public health and hygiene and innovations in advertising and marketing spawned a new attention to infant feeding. Increasingly, analyses of infant morbidity and mortality—in particular those launched by the burgeoning baby food industries, many of which had close links with scientists and medical people—pointed to the inadequacies and deficiencies of human milk. Simultaneously, the overall rise in infant survival—due to a range of factors—often got attributed to the increasing reliance on formula. As medical researchers used scientific findings to better explain and predict the course of disease, and as changes in infrastructures positively impacted public health, the public began to associate science and medicine with prestige and to regard them with confidence. As Rima Apple points out, " 'science' became practically synonymous with progress and reform" (17). Consequently, mothers began to see the medical professional as the expert, and to abandon advice from their own mothers, neighbors, and relatives—traditional advice—that had defined infant care in the past.

Janet Golden notes that pediatric books published between 1825 and 1850 began "to assert that the nursery was a medical domain as much as it was a domestic space" (53). Becoming "arbiters of wet-nurse selection and management," doctors during the second half of the century began to blame infant deaths on "inadequate nutrition, due either to deficient breastmilk or poor artificial food" (Apple 4). While breastmilk continued to be seen, for a while, as "best," it was regarded as the proper food only when it was in "proper condition"; many impediments precluded breastfeeding: an inactive breast, one that was "diseased, or lacking a nipple; scroffula, consumption, syphilis, puerile fever"; and a "fretful temperament or emotional upset" (Apple 6). Increasingly, though, investigators as a whole began to demonstrate "possible nutritional deficiencies in human milk" (72). Finding that the vitamin content of mother's milk varied from woman to woman, and that a particular woman's milk varied over time, many scientists saw this as an indication of breastmilk deficiency and began to administer dietary supplements of vitamins A and C, cod liver oil, orange juice, and various vegetables to avoid the possibility of rickets or scurvy (Apple 72).

Indeed, according to Apple, "Once their research had disclosed the variable nature of breastmilk, some physicians promoted artificial feeding with a food compounded of known ingredients in preference for the uncertainty of maternal nursing" (Apple 4). Others, concerned about overfeeding, underfeeding, and having control over the amount that babies were getting, touted the bottle, since "with the bottle you always know how much you have" (77). Dr. Rotch, writing in the *British Medical Journal* in 1902, and articulating a popular sentiment, claimed that "the mere fact of the milk being obtained from the human breast does not preclude many dangers which arise from it as a food, owing to the highly sensitive organisation of the mother allowing the mechanism of the mammary gland to be interfered with" (653, quoted in Palmer 203). He continues: "When this mechanism is interfered with good milk may also become a poison to the infant. . . . It is evident, therefore, that there is nothing ideal about breastmilk" (653). While we can attribute much of the hype around formula then to its status as something "modern" and "scientifically" calculated, another contributing factor, according to some perspectives, is sexism, and the fact that these substances were literally "man-made" instead of "woman made." Palmer implies the sexism emergent in many contemporary discourses when she notes that "the greatest problem with breastmilk was, in the minds of the doctors, that it came out of women's bodies" (203). Certainly, the familiar, age-old binary "man–culture; woman–nature," in which "culture," "civilization," technology, and science get privileged over things that are "natural," "earthy," bodily, and "of woman" plays a role here.

THE GROWING IMPACT OF ADVERTISING

An 1885 Mellin's Food advertisement from a childcare journal called *Babyhood*—headed "Advice to Mothers"—offers insight into the thinking of the period and the ways in which concern over disease entered into infant feeding discourse and behavior. It cautions "The swelling tide of infantile disease and mortality, resulting from injudicious feeding, the ignorant attempts to supply a substitute for human milk, can only be checked by enlightened parental care" ("Advice" quoted in Apple 106). Like many advertisements of the day, the format suggests that this is a piece from a medical treatise; and indeed, below the two-columns of texts, readers find that what they are reading is an extended quotation from a book by Dr. Harvey, *The Care and Feeding of Infants*, which, we are told "may be had free" by addressing Doliber, Goodale, and Co. As such, the text holds itself out as an objective, disinterested piece, rather than an advertisement. It claims: "Men of the highest scientific attainments of modern times, both physiologists and chemists, have devoted themselves to careful investigation and experiment in devising a suitable substitute for human milk." While the advertisement states that "if possible, mothers should nurse their children," it concedes and almost instructs that one's environment often prevents this: "the claims of society, artificial surrounding, and other causes, have been potent in promoting the use, if not the necessity, of artificial food." Increasingly, the nod to breastfeeding fades away from advertisements, parenting manuals, and mothering classes, so that by the early part of the twentieth century breastfeeding gets mentioned sporadically, if at all. We can witness similar advertising approaches today by Enfamil and Nestlé, acknowledging breastmilk, followed by the claim that each has now included in its mixture lipil, a blend of DHA and ARA, that renders them comparable to breastmilk.[8]

The 1885 Mellin's advertisement characterizes most in the period; Mellin's offers the best available through science. "A compound suitable for the infant's diet must be alkaline in reaction; must be rich in heat-producers, with a proper admixture of albuminoids of a readily digestible nature, together with the necessary salts and moisture" (v, quoted in Apple 106). Indeed, it boasts that "an analysis of Mellin's Food prepared with water and cows' milk according to the directions, shows the closest approximation to analyses of human milk." "Exactly suited to the ordinary powers of the babe's digestive organs," this formula gets packaged as the answer to fears and worries that have plagued parents: "by its use and the exercise of proper care those diseases which work such frightful havoc among infants—diarrhea, convulsions, the various wasting diseases, etc.—would be reduced to a minimum, and their fatal results be largely decreased." The ad continues, in the words of Dr. Hanaford, who makes a point noting that he practices in the vein of the empirical method: "This opinion is not based on theoretical assumption, but on observation and experience with the little ones I have attended . . . the babe will thrive on cows' milk and Mellin's Food alone, and will need no addition to its diet till about twelvemonths of age."

A 1907 advertisement for Mellin's—which presents twins, one "raised on Mellin's Food," the other "Nurtured at the Breast," in which Donald, the Mellin's baby, is larger and fatter than Dorothy, the breastfed one—suggests the seeming disingenuousness of claims that breastmilk—the only food available to poorer women—is better. The ad states: "We do not claim the Mellin's food and milk is better than mother's milk," yet the picture, assuming readers of *Good Housekeeping*, in which the ad appeared, will see bigger and fatter as better, suggests that the only option is Mellin's ("Mellin's Food," quoted in Apple 112). The gender division here, in which the boy gets the mathematical "formula" and the girl gets the "natural" and "clearly" less potent breastmilk, offers insight into the subtleties of the gendering that informs such discourses. In addition, if we read the pictures in the Mellin's advertisement through the filter of a 2001 study published in the *Journal of the American Medical Association* (JAMA) suggesting that while formula-fed babies may be bigger and fatter, this may not be a good thing, and linking formula-fed babies to a tendency to be overweight later in life and to complications associated with overweight status, we can see that infant feeding issues and advertisements warrant greater attention (Gillman et al.).

Increasingly, doctors and scientists infiltrated the domain of childbearing and childrearing. An 1899 *Ladies Home Journal* commentary states: "Ideal motherhood, you see, is the work, not of instinct, but of enlightened knowledge conscientiously acquired and carefully digested" (Apple 100). By 1911, this had become, in *Good Housekeeping*, "maternal instinct, left alone, succeeds in killing a large portion of the babies born into this world"; while the reference to "maternal instinct" is not clear, one can surmise that "mother's instinct" might be collapsed with "mother's milk." The growing focus on various formulas as the answer to infant disease and distress, and the increasing absence of any mention of breastmilk, in the aggregate, must have prompted many women to consider their breastmilk, if not as a "killer," at least as less salutary than formula. *Parents' Magazine*, in 1935, counsels: "Doctors, teachers, nutritionists and research workers are daily proving that not mother love alone, but mother love in combination with the best that science has to offer in all fields of childcare is needed" (Apple 96). Similarly, a 1938 advertisement for a course on the care and feeding of young children published in *Parents' Magazine* advises "instead of blindly following instinct alone or laboriously duplicating the tedious methods of previous generations . . . turn to specialists and authority" (Apple 99–100). Indeed, a range of authorities—the extremely popular "Well-baby Clinics" and "Little Mothers' Clubs," childcare manuals and columns in magazines, advertising campaigns of infant food manufacturers, and hospitals—all promoted "scientific" infant feeding and "scientific motherhood" (Apple 113).

MEDICINE AND MARKETS

Hospitals became one of the key sites for promoting artificial infant food as the answer to children's health. Increasingly, the hospital became the setting not only for the majority of births, but for the establishment of infant feeding practices. Whereas, in 1920, barely 20 percent of births in the United States took place in hospitals, by 1950, over 80 percent of women choose hospital births (Apple 159). The growing medicalization of childbirth along with the increasing medical interventions into the birthing process, with labor coming to be seen in terms of illness rather than a "natural" process to which women's bodies are suited, constitute another chapter in this story. As midwives became ousted from the birthing process, and the burgeoning (male-dominated) fields of obstetrics and gynecology took over what had been a woman's domain, launching a vigorous campaign against midwifery in the 1930s and eventually outlawing the practice, specialists also began to institutionalize postpartum arrangements—including the handling and feeding of infants after birth, until their departure from the hospital and beyond (Apple 173; Ehrenreich and English 31). Since nursing mothers were encouraged to sleep through the night and maintain strict feeding schedules, babies often cried and were given "supplemental feedings," which impeded lactation. That many hospitals held postpartum classes on how to mix and administer formula and sent mothers home with free samples suggests the degree to which the institutions promoted the bottle over the breast. Women were told that they "could be relieved of the burdens of nature through the wonders of modern science" and, Palmer notes, "many of them welcomed this liberation" (203).

We must recognize that the story involves other complexities—including the interplay of shifts in capitalist practices, market forces, and trade policies. And most studies that engage the question of the transition, in the United States and much of the rest of the world, toward artificial infant foods—particularly those that address what they see to be the callous and irresponsible practices of large infant food manufacturers in aggressively campaigning in "underdeveloped" regions where formula poses greater health risks—do point to the centrality of market forces. Palmer suggests that improvements in dairy production brought about surpluses of whey—the byproduct of milk pasteurization—and that these surpluses, in turn, prompted entrepreneurial spirits to search for ways to market the whey (203). Suggesting the degree to which many advocates view formula-use as having to do first and foremost with marketing, some surmise that formula developers use whey as the base for artificial baby milk not because research proved it to be the best, but because it presented itself through the vagaries of our particular market economy as a surplus good that needed a use (Palmer 203). On the other hand, we can surmise from surveying the literature available that many who developed formulas did so out of a concern over high infant mortality (Apple 111). Apple notes that the reasons for the switch are complex, and that "manufacturers, though sometimes developing their products for humanitarian reasons, needed to build consumer demand for infant foods" (169). Several scholars suggest that doctors remained reluctant to promote infant formula well into the 1920s, but that by then the use of commercial foods had become so well established that they could not stem the tide of support.

For many of these scholars, the marketing by commercial industry and promotion by doctors became difficult to separate (Apple; Palmer 207). Apple suggests that, by the turn of the century, most doctors tended to promote artificial feeding, holding themselves out as the authority on feeding methods. In this light she quotes an editorial from *Hygeia*, a popular health journal published by the American Medical Association at the time. It recommends to consumers that they "read the advertisements as [they] do the articles in the magazine"—many of which promoted various formulas— since they are "packed with facts" (quoted in Apple 118). Palmer also points to the growing "medical/commercial liaison," and many sources corroborate this link (Apple; Baumslag and Michels 147).

On the question of health, it is important to note that nineteenth-century physicians drew links between high infant mortality rates—specifically those from gastrointestinal diseases—and poor nutrition, often the result of bad artificial feeding (Apple 109). Several scholars note that the limited statistics available give credence to physicians' concerns about this relationship and to their assertion that improved artificial feeding in some instances reduced the infant death rate (Apple 171). In the 1890s, the mortality rate from diarrheal diseases declined in many places after the establishment of Straus-type milk depots in cities, which helped to foster safer cow's milk, often used as the basis for infant food. Similarly, early twentieth-century pediatric textbooks often point to the precipitous drop in the mortality rate of children from the late nineteenth century onward. Noting a sharp fall in deaths from gastrointestinal diseases, physicians claimed that this reduction resulted from "improved milk supplies, better maternal education in infant feeding and hygiene, and increased medical supervision of infants" (Apple 170). Over time, formula came to be linked, by many, with improvements in child morbidity and mortality. While some see the connection as constituting an "open question," it is clear that the increasing availability of safer milk and healthier formulas meant that a smaller proportion of bottle-fed infants died than had in earlier days of formula feeding (Apple 171). We also know that mothers tended to supplement with water, and that improvements in water supplies "certainly contributed to a decline in gastrointestinal diseases," while other improvements in living conditions undoubtedly had a positive impact on infant health overall (Apple 171,3). Feminist anthropologist Van Esterick suggests that while pediatricians, early on, recognized the dangers of feeding infants artificial infant food, these dangers had the effect of increasing the need for medical supervision of infant feeding, ultimately strengthening the relationship between industry and the medical profession (*Beyond* 118).

What is interesting is that while we know of many problems with formula contamination during this period, especially during the early years when sanitation and transportation issues rendered the use of cow's milk dangerous, when lack of infrastructure made for questionable water supplies, and when formulas themselves contained some problematic ingredients, this did not seem to garner much attention, yet breastmilk contamination, particularly in the mid-nineteenth century, and as it related to poor and immigrant women who served as wet nurses, seemed to matter greatly. According to Shaftel's study, "History of the Purification of Milk in New York," "There was little popular awareness at the time of the relationship between the quality of the [cow's] milk and the epidemics of diseases which were so devastating, particularly to the infant population" (277). But, on the other hand, his research also suggests that popular concern over disease did not always follow from awareness of it. His recounting of the story of swill milk in New York city in the period between the late 1830s and the early twentieth century indicates the degree to which even extremely dangerous practices, when they are supported by powerful interests, sometimes go undetected for quite some time without the public's becoming alarmed.

"Swill milk" refers to the milk produced by feeding cows the waste or "slop" left over from distillery activities; once someone discovered that feeding cows this mash produced, at least for a while, larger amounts of milk, dairies were built next to distilleries so that troughs led directly from the still to the mangers (Shaftel 275). After starving cows for a period, they could be induced to eat the distillery slop—this despite the fact that their milk, when boiled, would "smell strongly of beer and would coagulate into hard lumps" (277). Robert Hartley and others campaigned against the practice, which produced milk dangerous to babies, especially at a time when mothers were increasingly employing formulas that required mixing with cow's milk. First appealing to the press in 1836 and 1837 and receiving little support, Hartley published an expose in 1842 that vividly depicted the process of "gorging" animals "so inhumanly condemned to subsist on this most unnatural and disgusting food" ("An Historical" quoted in Shaftel 278). In his pamphlets, he describes a typical stall housing 2,000 cows, in which "the cattle, head to tail, stood in rows about three feet wide which would permit the cows no movement, and none there was for an entire nine month, during which they would be milked while standing, weak and sicly, up to their bellies in filth and excrement" (Shaftel 278). But, even as the dangers to babies were revealed in the most vivid language—with Hartley claiming that the high infant mortality in New York "was largely the result of the milk from these cows"—the public remained, in large part, unconcerned. Finally, in 1848, the New York Academy of Medicine appointed a committee to investigate swill milk; it corroborated what seemed apparent: they found that children fed on it were susceptible to scrofula and cholera as well as to whatever epidemic disease manifested itself at any given time (279). Despite renewed attention in the *Sunday Dispatch* in the early 1850s, it was not until 1858, when the "roisterous cartoons and journalistic sensationalism" of the *Illustrated Weekly* called attention to the issue, that the Common Council of Manhattan appointed a committee of aldermen to look into the problem, at which time the aldermen were won over by gifts from the distillery owners (280). After all, besides offering a solution to the problem of cows living in situations where feed had to be brought from the countryside, distillery slop milk proved financially profitable as it offered, at least initially, greater milk yields (Shaftel 277).

The delay in bringing sufficient attention to this issue so that change would occur is instructive. It was not until 1862 that the first milk law was passed in New York state; it forbade "the feeding of cows on food which would produce unwholesome milk"(281). Nevertheless, in 1904, one could find industries in New York City in which over 6,000 cows continued to be fed distillery slop. To understand the danger of such milk, we can look at a 1910 study of New York City that showed that although more than 75 percent of babies were breastfed, 78 percent of the infants who died of enteritis were fed either cow's milk or patented formulas (Shaftel 285). One study indicates that infant mortality increased from 32 percent in 1814 to 50 percent in 1841; these figures "coincided with those of other cities where milk was similarly produced." In contrast, in European cities, where the use of swill milk had not been established, mortality rates decreased in the same period (Shaftel 277). The comparison is significant when we analyze comparisons between the breastmilk of women here and that of women from countries who have imposed stricter legislation banning certain chemicals. When we examine infant formulas, we see that while many got more advanced and safer over time, problems have existed in many periods. For example, while all advertisements of formula in the early years claimed "purity and sterility," a 1911 study in Britain discovered every brand to contain microbes (Palmer 194). Such findings, taken with the lack of alarm over problems with artificial foods and the anxiety over the spread of disease through wet nurses, suggest, as noted earlier, the degree to which fear over infant food contamination does not always match the risk involved. Instead, it seems, at least in these contexts, to reflect other anxieties associated with social, demographic, economic, and other changes that left people feeling vulnerable and open to discourses of certainty—such as those increasingly offered by a medical industry proclaiming the healthiness of various mathematically rendered formulas.

COLONIZATION AND THE SEARCH FOR NEW MARKETS: EXPORTING INFANT FOOD AS THE MORE "CIVILIZED" CHOICE

In moving from the era in which formulas began to be widely used in the United States and Britain, to periods, not too much later, when they began to be exported to "developing" countries, a few broad trends emerge. If we turn to the waves of colonizing endeavors that mark much of recent history and continue today in varied forms, including the search for cheap labor and raw materials, we find that colonization has had dramatic impacts on infant feeding and infant health worldwide. As Palmer notes, we tend to think of the places colonized through the imperialistic endeavors of Europe and the United States as providing either raw materials for "first world" use or cheap labor for "first world" markets (199–227). But another important role played by the colonies and former colonies, especially in the twentieth century and beyond, has been that of providing markets for "first-world" goods. The "developing" world has become a huge market for infant formulas, a $1.5-billion industry in the United States.[9] As with other colonizing influences, practices employed by the first world often get translated to the colonized; breastfeeding trends in many parts of the developing world have followed trends in the developed world, so that rates have been very low in many parts of the world over the last half-century, and still remain low in many places.[10]

Much of the debate over infant feeding has centered on the use of formula in areas with poor water, poor infrastructures, illiteracy, and a range of other circumstances that can and often does harm infant and child health, as well as families. Formula-use in many areas means that babies get subjected to a range of health risks. When we consider the advent of the age of formula-fed babies in developed countries, questions about disease and contamination clearly need to enter the discussion. But when we consider the advent of the age of formula-fed babies in developing countries, there is no question that disease and contamination loom large. Water treatment facilities tend to be much less prevalent in many regions of the developing world, so that the availability of clean water to mix with formula and use for cleaning bottles and nipples is often limited. In addition, in tropical climates, uncontaminated milk and formula are much more susceptible to contamination than in cooler climates, so that feeding infants formula in these zones poses much greater risks. And, of course, economics plays a role here. In poor countries where women receive free samples of formula, lose their breastmilk supply, and then cannot afford to purchase sufficient formula, these women sometimes have to dilute powders with additional water, often depriving their babies of adequate nutrition. The economic impact of having to purchase formula in contexts where many families are stressed economically means that additional burdens get placed on women, who are shown in many studies to bear the brunt of economic stress before men (Vickers 9).

When poorer women and women in poorer countries begin to use formula, the relative expense of continuing the practice is clearly a greater burden for those families than it would be for the families of women of greater economic means. Not only must we weigh the cost of the formula itself, but we must also take into account the cost of health interventions to address illnesses linked to the mixing of formula with unsafe water. In addition, we must calculate the costs of medical interventions for health issues associated with lowered immunological functioning as a result of using formula. Palmer points to a 1924 study of 20,000 infants from poor families in

Chicago that found that at nine months of age the formula-fed infants were fifty times more likely to die than those infants who were still breastfeeding (210). Eventually, better living standards, education, medical care, the availability of immunizations, antibiotics, and other medications in wealthier regions offset many of the more dangerous effects associated with artificial feeding in more developed regions. But in less developed areas, one can still witness grave problems with infant formulas. For example, a 1985 study of an urban area of southern Brazil showed that bottle-fed babies were four times more likely to die of respiratory infection and fourteen times more likely to die from diarrhea than those exclusively breastfed, thus suggesting that, in many regions, infant formula presents ongoing problems (Cesar 319–22, noted in Palmer 210). Many breastfeeding advocates blame the colonizing endeavors of formula manufacturers for the sometimes dire impacts on babies' health, characterizing companies as valuing profit over life.

If we want to read the situation in a less cynical way, it makes sense to recognize that as the tide turned toward using formula in Britain and the United States—arguably among the most dominant and dominating countries throughout much of the nineteenth, twentieth, and twenty-first centuries—many promoters of artificial food abroad, at least in the early years, increasingly came to argue, if not to believe, that they were offering their "less privileged" counterparts something possessing marked benefits over breastfeeding. As science, technology, and the importance of doctors, hospitals, and formula to childbirth and childcare came to define practices in the colonizing countries, so too did these zones of influence begin to infiltrate practices in the colonized countries (Palmer 208). Indeed, the medical/commercial relationship that defined childrearing for many years became a model for similar relationships throughout the world (Palmer 208). If things "scientific," "civilized," and "technologically advanced" held appeal for American and British consumers, some historians argue, this appeal often took on an added dimension for peoples subjected and suppressed by being denied the ability to govern themselves. According to Palmer, "the fact that [formula] was an imported product and was already used by the colonial elite" in many regions "added to its status" (230). Also of importance is the fact that in many cases, for a long expanse of years and in a range of settings, a host of organizations—UNICEF, charities, church missions, and relief agencies—all promoted formula in the belief that they were offering good advice. And breastfeeding advocates continue to blame all formula promotion, linking it to larger issues of control. For example, Palmer states that, as time went on, many agencies "did untold damage to breastfeeding and to the economic and health independence of newly independent countries through energetic milk promotion" (226). Several scholars note that such practices continue today in refugee camps and other loci of international aid, so that the dependence of poor nations on formulas purchased from huge corporate interests continues to be at stake (Palmer; Van Esterick).

It bears repeating that the availability of formula has also saved lives. When large numbers of mothers have died of epidemics, for example, formula has meant the difference between life and death. But, on the other hand, the degree and nature of marketing practices suggest that, in many cases, what is at stake is market expansion, not saving babies' lives. Historians and advocates point out that when doctors and health clinics in many areas of the developing world have offered free samples of formula—sometimes as a lure to get women to take advantage of beneficial medical treatments and immunizations, and sometimes simply to promote the habit of using formula so that the option of nursing becomes less available—many indigenous women have abandoned more traditional approaches in favor of the more "civilized." Palmer notes "In many societies any non-human milk is perceived as a replacement for breastmilk and donation is interpreted as a health message to give this to a baby instead of suckling" (226). While representatives of the formula industry and colonizing medical practitioners certainly played a role in bringing about a trend toward the use of formula in many regions, it should also be noted that, in many cases, the formula industry, acting alone, became the disseminator of the practice. Several scholars have pointed to situations in the past in which formula companies dressed their employees in doctors' and nurses' clothes and, posing as medical practitioners, prescribed formula as the best food for infants (Baumslag and Michels 147–50; Palmer 229).

Infant food and marketing it to developing nations continue to be a hotly contested arena in assessing current globalization trends. Another less discussed impact involves questions around toxicity both abroad and within the United States. As I will discuss in chapter two, over time, colonizing activities set up conditions for increasingly low breastfeeding rates worldwide and subsequently less attention to breastmilk toxicity. In addition, that the Nestlé boycott and other campaigns around formula marketing practices in developing nations have long dominated our attention means that notice gets drawn away not only from breastmilk contamination issues, but from marketing issues and policy issues within the United States, and from ways in which global trends and local trends diverge and interlock. Undoubtedly, millions if not billions of babies have been impacted by how we tend to frame contamination issues primarily in terms of formula and effects on babies in the developing world.

The problem of whether limits should be placed on the direct marketing of infant formula has a long, intricate history; due to the efforts of countless breastfeeding advocates, many countries have signed a code in which they agree to restrict the direct marketing of formula to women and to hospitals.[11] While advocates continue to note numerous infractions of this code, with negative health outcomes for babies, including greater subjectivity to various diseases and other contamination issues, they also continue the work to confront ramifications for people living in countries that have not signed on to the code, including the United States. As a key player in the world scene, and a country whose influence spreads to other countries, the United States is also important in that its practices do tend to get disseminated. Breastfeeding advocates point to how the U.S. government's tendency to favor industry has resulted in the deprivileging of infant health in many areas. In Sweden, where breastfeeding rates used to be roughly the same as in the United States, women now nurse at the highest levels worldwide, with a 98 percent initiation rate and 80 percent still nursing at six months (Phillips et al. 587). That Sweden signed the international code against direct marketing and launched a vigorous campaign in support of breastfeeding implies the importance of the code and offers a model for how things could be handled in areas where breastfeeding rates remain low, including the United States.

BREASTFEEDING AND POPULATION DEBATES

While infrastructure issues, lack of clean water, poverty, and other conditions affect formula-use in third-world contexts, with healthcare workers, nongovernment organizations (NGOs), relief agencies, and breastfeeding advocates pointing to a range of negative effects on infant health, some of these advocates sometimes broach the issue of lactation's effect on ovulation. By delaying ovulation, breastfeeding limits a woman's fertility and so helps to ensure that her births get spaced in such a way that she is less likely to suffer the effects of too frequent and too closely spaced pregnancies. Although frequent pregnancies can strain maternal and child health in favorable contexts, they can be devastating in contexts in which women's health and access to food and healthcare may be compromised. While these factors play an important role in our thinking about infant feeding, and affect the health and well-being of many, it is important that we address the complex web of issues surrounding maternal and child health, both abroad and within the United States. Advocating for breastfeeding in light of its effect on fertility and the consequent benefits to infant and maternal health is useful, but it is important to recognize that discourses around fertility have a long and problematic history. Sensitivity to this history should be acknowledged in discussions about infant feeding, especially in cases when environmentalists, NGO's, international agencies, and governmental agencies play a role.

I want to turn now to 2003, which was dubbed the "year of the 6 billionth child" (Y6B). Discussions about population and habitat arose out of the growing attention to numbers. Indeed, groups of people met to confront the ramifications of burgeoning population growth, some of them concerned with effects on shrinking biodiversity, declining animal habitat, and myriad impacts of the increasing destruction or sullying of the earth's resources. Many groups, including coalitions addressing the need for women's access to healthcare and voluntary family planning, groups concerned with the need for international aid programs to be funded at higher rates, particularly in

light of the growing gap between rich and poor nations, and groups addressing impacts on a range of ecological systems, came together to search for new
approaches. The controversies that arose out of this attention to Y6B illustrate some of the tensions that result when fertility issues, particularly for
women in the developing world, get invoked. While some groups focused on
complex webs of occurrences, and addressed issues with sensitivity and responsibility, some focused on numbers and the need to limit certain women's
fertility, particularly those from developing nations.

Claims emerged that some groups, in urging attention to population
growth and fertility, were promoting "fertility control" or engaging in acts of
supremacy and colonization. For example, controversy sprang up around a
special town hall meeting in my hometown of Albany, New York. A group
called the Committee on Women, Population, and the Environment (CWPE),
"an alliance of feminist activists and scholars" who support "women's rights
to safe, voluntary birth control and abortion, while strongly opposing demo-
graphically driven population policies," responded to the meeting by taking
issue with the very premise of the meeting. They issued a flyer in response,
asking people to boycott the event and to "Stop Demographic Alarmism!!!"
and "Support Women's Rights, Not Population Control!!!" (Flyer February
2002). The open forum meeting, which was carefully planned by a range of
groups including representatives from the Upper Hudson Valley Planned
Parenthood, the National Wildlife Federation, the United Nations Population Fund (UNFPA), and SUNY's Women's Studies Department, had a clear
mission of recognizing the need for sensitivity and awareness of complexities,
including the recognition of the importance of the move away from population control and coercive strategies toward providing healthcare, education, and choice for women and girls. It rejected the tactics of coercion and
fear mongering and addressed the necessity of confronting vast disparities in
consumption and waste production between first- and third world nations.
Conference organizers pointed to our own high consumption rates and the
heavy generation of waste products and environmental pollutants that result
from high industrialization as being key causes of the limits on earth's carrying capacity. Nevertheless, the CWPE accused conference organizers of
insensitivity and suggested that their approach to the problem of Y6B
amounted to an extension of earlier promotions of coercive limits on the
fertility of some, but not others. This meant that some of the more balanced
calls for multipronged approaches that were being promoted were dismissed.
And it meant that those who heeded the flyers and boycotted the event
would be deprived of an opportunity to dicuss the issues and share their
reservations. On the other hand, the situation did provide an opportunity to
examine CWPE's concerns.

Discussions around population issues and women's and children's access to healthcare involve infant feeding issues to the degree that infant
feeding choices impact women's fertility and so become part of the subtext

for discussions about population, consumption, habitat destruction, and the need to protect earth's resources from further degradation. Any discussion engaging the importance of breastfeeding for women in the developing world that invokes the limits on these women's fertility or addresses certain women in first-world contexts who are perceived to be "overly fertile" must involve careful attention to the history of coercive approaches to women's fertility. As Angela Davis, Andy Smith, and several other scholars have suggested, to invoke third-world women's fertility or the fertility of certain groups of women as "the problem" is to step into a long history of colonizing and racially charged approaches that have served to alienate many women, leaving them vulnerable to further scapegoating. Angela Davis notes, for example, that even though the "progressive potential of birth control remains indisputable . . . arguments advanced by some birth control advocates have sometimes been based on blatantly racist premises" (202). Andy Smith focuses specifically on Native American women. She points out that even to suggest that checking these women's fertility constitutes a solution is to participate in oppressive patterns. And she chides the Sierra Club and Planned Parenthood for implying that overpopulation leads to environmental deterioration. Smith claims that such approaches draw attention away from other occurrences that place greater burdens on the earth. In the words of Betsy Hartmann, these include "dominant economic systems which squander natural and human resources in the drive for short-term profits; and the displacement of peasant farmers and indigenous peoples by agribusiness, timber, mining and energy firms" (91–92 quoted in A. Smith 27). They also include "the role of international lending institutions, war and arms production, and the wasteful consumption patterns of industrialized countries and wealthy elites the world over in creating and exacerbating environmental destruction." Many who have advocated for greater funding for international agencies and NGOs involved with providing women worldwide with access to basic healthcare and reproductive care have had to learn some hard lessons about the delicacy of the situation.

That the 1994 Cairo conference on population changed its strategy to focus, not on numbers, but on providing women and girls worldwide with greater access to healthcare and education suggests that the work of those calling attention to supremacist and racist gestures has begun to pay off. But the issue still remains a challenge. That environmentalists have sometimes promoted addressing population growth in order to limit encroachments on animal habitat or on earth's fragile ecosystems has meant that they have been vulnerable, at times, to rhetoric accusing them of colonizing gestures. Such discourses affect how discussions of infant feeding in various contexts play out, and so warrant mention here. It is unfortunate that when some environmentalists or breastfeeding advocates have addressed the fertility-limiting properties of breastfeeding, particularly for women in the developing world, they have thus left themselves open to accusations of supremacy and imperialism, but that when the most conservative fringes of the right, in their hatred of any policy that they construe as being pro-abortion, make sweeping policy changes, their actions are so predictable that they sometimes garner less attention. Changes in some groups' approaches to this issue suggest that several years in which accusations have been hurled around have begun to pay off. For example, discussions of population issues on the websites of key environmental groups today exhibit much more sensitivity than in previous years. The Sierra Club, for example, begins its discussion of global population and the environment noting that it is consumption and degradation of natural resources that "jeopardize the health of the planet" ("Population"). While earlier websites paid less attention to the wording of population issues, today's website is careful to emphasize support of "voluntary family planning" and the "rights of women and girls to education." Foregrounding consumption issues, Sierra notes that "with less than 5 percent of global population, the United States accounts for about one-fourth of global consumption, so that a child born in a developed country will consume and pollute more over his or her lifetime than thirty to fifty children born in low income countries" (Denny et al. quoted in "Population").

DEBATES AROUND THE POSSIBILITY OF HIV TRANSMISSION THROUGH BREASTMILK: A NEW CHAPTER OR AN OLD STORY?

While infant feeding in relation to population issues continues to spark controversy, debates around the probability of human immunodeficiency virus (HIV) transmission through breastmilk have become surprisingly complex. And, as one might expect, these are often hotly contested. That such debates involve significant aspects of the history laid out in this chapter suggests the degree to which child nourishment issues continue to serve as a template for reading how our approaches to this seemingly insignificant event—choosing how to feed one's child —mirror larger cultural phenomena. Part of the controversy stems from the fact that relatively little research has been done, and that it offers mixed results, and part of it has to do with the reality that some have sought to make recommendations for all women in all situations, regardless of the set of circumstances faced by individual women and babies. I do not believe these are intentional moves calculated to harm women and babies; rather they reflect the difficulty researchers and policymakers face. In some cases, attempts to dictate what various women should do have brought about what amounts to colonizing gestures and led to charges that formula companies are capitalizing on the issue and holding themselves out as the saviors of babies or that breastfeeding advocates cling to ideologically rigid positions that harm babies. Accusations that babies are being left vulnerable fly in all directions.

Results suggest that, with no treatment, approximately 25 percent of infants born to HIV-infected mothers will become infected (Callahan,

Caughey, Heffner 81).[12] Because researchers believe that transmission occurs late in pregnancy or during labor and delivery, particularly through exposure to maternal blood, doctors have, in some cases, opted for caesarean delivery, which has been shown to lower transmission rates by two-thirds. In addition, current evidence suggests that zidovudine (ZDV) or AZT administration during the antipartum (after the first trimester), intrapartum, and neonatal periods can reduce the risk of maternal-fetal transmission by two-thirds in women with mildly symptomatic HIV disease (81). But many, many women do not have access to these treatments. When we get to the question of whether an HIV-positive mother should breastfeed her baby, the standard approach, here in the United States, involves the recommendation that infected mothers not breastfeed their babies, as the virus has been found in some percentage of cases—usually around 16 percent—to be spread through the mother's milk (Callahan et al. 81).[13] But the issue is much more complex than those numbers suggest.

An article appearing on WebMD Health, an online source that physicians, medical students, other health practitioners, and laypeople consult for "standard" information, highlights some of the controversy. It quotes Dr. Paolo Miotti, medical officer, divisions of AIDS, National Institute of Allergy and Infectious Diseases (NIAID), and lead author of a study appearing in the 1999 *Journal of the American Medical Association* (*JAMA*). Explaining the results of a three-year study that was conducted in Malawi, Miotti points out: "One recommendation to stop the transmission of HIV to breastfeeding infants internationally would be: Don't breastfeed if you are infected with HIV."[14] While highly recommending that HIV-infected mothers here in the United States not breastfeed, the article suggests that some confusion surrounds the question of the merits of HIV-positive mothers nursing in less developed countries. On the one hand, the article seems to advocate for the position that *all* HIV-positive women, regardless of location and circumstances, should desist from nursing. It notes that "the study's findings are likely to have a strong impact on international recommendations for limiting the spread of the disease," since, according to Miotti, breastfeeding "is almost universal" in developing countries. On the other hand, it gestures toward the position that breastfeeding advocates have fought hard to instill in those making recommendations for third-world women. It notes, for example, that "in 1998, the Joint United Nations Programme on HIV/AIDS issued a revised statement suggesting (1) that women be offered HIV testing and counseling, (2) that they be notified of the benefits and risks of breastfeeding if the mother is HIV-infected, and (3) that they make a decision [about breastfeeding] based on individual and family situations" (Kuhn).

Conceding that "bottle feeding is a very expensive and impractical solution" in many countries where "prepared infant formula also may be contaminated by local water supplies," Miotti notes that "an editorial in the same issue of *JAMA* points out that the risk of HIV transmission may decline over the course of breastfeeding."[15] Indeed, that editorial, by Fowler, Bertolli, and Nieburg, recognizes that many women in resource-poor countries risk social stigmatization of themselves and their infants, including violence and abandonment, if they choose not to breastfeed, that many women lack access to testing, and that "there is little current information to provide an HIV-infected women regarding her infant's chances of dying from diarrhea or respiratory diseases if she chooses not to breastfeed" (782; "African Breastfeeding"). Together the articles suggest the degree to which controversies continue to surround the issue.

While many point to the increased dangers of formula-use for women in areas with weak infrastructures, lack of access to clean water, and minimal healthcare, these same concerns arise when the question of HIV-contaminated milk gets invoked. Formula-use for babies in these situations can be dangerous and sometimes deadly. Diarrhea and other diseases caused by contaminated water used to make formula or through the denying of breastmilk that could help boost the babies' immune system prove a big problem for babies in these locations (Van de Perre 122).[16] In many locations with large segments of the population unsure of their HIV status combined with the limited access to clean water and healthcare, many advocate breastfeeding because denying the immunological properties of breastmilk to such babies is seen as risky (Babymilkaction Update 25; "HIV and Infant").

"African Breastfeeding Rates Threatened," an article published by InfactCanada, an advocacy group, notes that governments, UN agencies, and others, in response to "concerns about declining action to protect, support and promote breastfeeding in Sub-Saharan Africa," planned a fact-finding visit to four countries. Noting that the decline in breastfeeding support is "related to the HIV/AIDs pandemic," the group explains: "The concern that mothers may infect their infants through breastfeeding has had a massive spillover effect, in which concerns for HIV/AIDs has been transferred into a decline in breastfeeding support." They warn that "if the concerns about mother to child transmission are translated into heavily funded efforts to convince mothers to avoid all breastfeeding, either because they are HIV positive, or believe they may be HIV positive, the result will be disastrous for breastfeeding in Sub-Saharan Africa." Insisting that context matters greatly, breastfeeding advocates contend that the percentage of infants infected through breastmilk is relatively small compared to the total numbers of infants in Sub-Saharan Africa, and other sensitive contexts, who would be adversely effected by a decline in breastfeeding. UNICEF estimates that in the last twenty years as many as 1.7 million children have contracted HIV through breastfeeding, and that during this same period 30 million children have died who would not have, if they had been breastfed (Martyn).

Feminists and social justice activists often make the case that the most responsible approach involves allowing those affected to speak for themselves. And we should recognize that child health specialists in South Africa and other areas are urging governments and agencies not to provide free milk in programs concerned with preventing mother-to-child transmission ("Free Formula"). Specialists from the University of Natal and the Child Health Group of the Africa Centre for Population Studies and Reproductive Health report that "free formula may appear a blessing, but while potentially decreasing the rate of postnatal transmission, it is very likely to increase morbidity and mortality from other infectious diseases, thus decreasing overall child survival" (Coutsoudis 154). A study from London "Preventing mother-to-child transmission of HIV-1 in Africa in the Year 2000," hints at the limits of making recommendations for women in developing countries based upon assumptions that more fully fit first-world contexts. Its suggestion that "a short regimen of peripartum ARV with alternatives to breastfeeding . . . from birth currently represent the best option to reduce MTCT of HIV in Africa" fails to take into account the conditions specific to various African contexts (Dabis et al.; Nduati et al.). In addition, it is important for those involved to recognize that many factors, including sexism, colonialism, and racism, determine which mothers and babies enjoy the right to various options, including choices about infant feeding, drug therapies and their availability, and even choices about how to protect themselves and their babies from further harm.

The controversies remain. Breastfeeding advocates point out that the link between HIV and breastfeeding opened up an opportunity for formula companies to reposition themselves as "saviors" rather than "culprits" in the debates around infant sustenance. They complain that "artificial baby milk companies (and drug and other companies who have a vested interest in promoting the HIV 'market') have been offering donations of products and services to NGOs to use in programmes to reduce transmission, and at the same time lobbying governments to weaken legislation concerning the marketing of breastmilk substitutes" (Martyn). As an example, they note that the International Association of Infant Food Manufacturers (IFM), the umbrella organization for the artificial baby milk companies, "told delegates at the World Health Assembly that mothers could not breastfeed if they tested positive for HIV" and cited UNICEF figures of comparative risks.[17] And they also note the excited tone of claims made by formula companies that HIV has opened up a vast new market. Bristol-Myers, for example, notes on its website that "the returns will ultimately materialize . . . most of this HIV market is untapped" (Bristol-Myers Squibb Company). Another piece of the advocates' argument involves changes in formula, and the fact that "the very same companies who are trying to promote their products as the solution to HIV transmission" are the same ones "who have taken out patents on certain components of breastmilk, such as lactoferrin" because "it is known to have anti-viral properties, which denature HIV" (Martyn). Baby Milk Action points out that "perhaps it is no longer correct (if it was ever) to say 'breastfeeding transmits the virus'," and suggests that "we should now say 'for infants of mothers diagnosed HIV positive exclusive breastfeeding can offer as much protection, maybe more, than artificial feeding' " (Martyn). In the effort to suggest that their chief concerns is to ensure that mothers get "objective and sound information," these advocates note that the WHO International Code of Marketing of Breastmilk Substitutes allows for formula to be used as and when appropriate (Martyn).

The United Nations (UN) supports exclusive breastfeeding when conditions for formula-use cannot be assured. Findings that some governments and NGOs have been distributing free formula to all infected women, regardless of other variables, has prompted claims that such practices are negatively impacting entire populations. The controversy has gotten fervent at times; for example, Babymilk Action points out that "*The Wall Street Journal*, followed by many other papers, made a vicious attack on UNICEF over its refusal of donations of formula for the UN pilot projects on the reduction of HIV through mother to child transmission" ("HIV and Breastfeeding").[18] Infactcanada notes that the authors of the piece "true to their Wall Street roots, promoted formula makers Wyeth and Nestlé as standing ready to save African babies with donated formulas from the scourge of HIV-infected breastmilk." And UNICEF gets presented as holding an "ideologically driven stance against the industry" coupled with a drive to "insist on compliance" with the International Code—here represented as not allowing that formula-use is recommended when appropriate ("Breastmilk Attacks").[19]

Stories of coerciveness have surfaced, with mothers reporting having been offered free antivirals only on condition that they agree to desist from breastfeeding, and with the Nambian government having to distance itself from a Bristol-Myers initiative in favor of its own "multi-sectorial National AIDS cordination programme" ("Philanthropy Online"; "President of Botswana"). That women from many developing nations have been targeted for coercive practices in other situations, along with the fact that evidence suggests a willingness on the part of formula industries to capitalize on the situation, brings the level of tension up a notch. The controversy has led the World Health Assembly Resolution 54.2 to call on member states to "assess the available scientific evidence on the balance of risk of HIV transmission through breastfeeding compared with the risk of not breastfeeding" ("Nestlé Uses"). But reports that Nestlé has claimed that its Nan Pelargon formula will "kill germs" in unsafe water serve to cloud the issue further and raise the stakes and tone of the debate ("Nestlé Uses"). One feature of the debates concerns the issue that richer producer nations are seen by many advocates as lobbying for weak measures governing standards and labeling in poorer nations including India, Brazil, Indonesia, Tanzania, Kenya, Romania, Bulgaria, and Bolivia ("Health Claims"). Given the history of other coercive practices and policies affecting health outcomes in poorer countries, this kind of claim has significant consequences.

Comments by Dr. P. K. Dlamini, Minister of Health in Swaziland, suggest the inappropriateness and danger of representatives of industrialized countries—whether healthcare workers, agency administrators, or formula industry liasons—making decisions for women in developing countries. Articulating a position that many feminists have championed in recent years, she concludes that "the best way is to provide women, the families, with all the information at hand at the moment and allow them to make informed decisions" ("Swazi Health"). The WHO, UNAIDS, and UNICEF Guidelines on HIV and infant feeding seem to do just that, suggesting "a number of infant feeding options," and noting that "the infants nutritional requirements, risks of contamination when formulas are used, increased risk of death due to diarrhea, as well as costs of artificial feeding, losses in fertility protection, the loss of optimal mother–infant bonding and the social cultural patterns of breastfeeding need to be taken into account" ("HIV and Infant"). They also note that "if a mother is infected with HIV, it may be preferable to replace breastmilk to reduce the risk of HIV transmission to her infant."

While the issues at stake for people in developing nations are complex, involving a range of factors, those at stake when we turn to first-world contexts add to the long list of controversies. With research suggesting that in first world contexts it may be best to recommend that HIV-positive women desist from breastfeeding, the question that emerges is: should a woman be required by law to desist breastfeeding if she tests HIV positive? This controversy ocassionally becomes heated. A group called MOMM, Mothers Opposing Mandatory Medicine, notes that "due to the questionable accuracy of HIV tests and the toxic and experimental nature of AIDS drug treatments, parents affected by these issues need information from a variety of viewpoints and authorities upon which to base crucial decisions and choices" ("Who Are"). La Leche League International's media release on "Breastfeeding and HIV" issued July 4, 2001, acknowledges the "worldwide challenge of making informed infant feeding decisions when HIV transmission is a consideration" ("Breastfeeding and HIV"). Noting that "for women who know they are HIV positive and where infant mortality is high, exclusive breastfeeding may result in fewer infant deaths than feeding breastmilk substitutes and remains the preferred feeding approach," they concede that things may play out differently in geographies where infant mortality is low. While maintaining that "there is no clear, published evidence that feeding breast milk substitutes [in these locations] results in lower infant morbidity and mortality in any infants," and asking us to consider the "social costs" of not breastfeeding in these situations—of which there are many—they urge "ongoing support of exclusive breastfeeding," and ask for "research studies that fully define the role of breastfeeding patterns," particularly studies that consider "exclusive breastfeeding and optimal breastfeeding management."

In light of the lack of clear evidence, given that many published studies do not focus on exclusive breastfeeding, which has been shown to have significant benefits, and pointing to the "inconclusive nature of the research and its various interpretations, the group states: "LLLI is not making a recommendation about breastfeeding for HIV positive mothers at this time," even while it laments that "fear of infections of HIV in breastmilk is . . . the most serious new threat to the practice of breastfeeding around the world," with "an unknown impact on many babies" ("Breastfeeding and HIV"). Attesting to the importance of remembering the long history of coercive practices, including those surrounding breastmilk contamination through the years, it notes that "modern medicine"—which has from time to time been linked both with the formula industry and colonizing approaches to health—"has a long history of hostility to breastfeeding and other natural practices." It suggests, therefore, that we remain cautious of facile mandates that all women, regardless of their situation, who test positive for HIV should use breastmilk substitutes. In other words, the history that I have laid out in this chapter continues to have a huge impact on how infant feeding decisions get made.

Accounts of coercive practices, including cases in which women's children have been taken from them, sometimes by court order because they refused to stop nursing, suggest that controversies for first-world contexts sometimes rival those in the developing world.[20] Instances in which governments have refused to allow newborn infants of HIV-positive women to be breastfed, when babies get exposed to various drugs through breastmilk— whether recreational or pharmacalogical drugs—and in which governments impose different standards on different women, sometimes reflecting classist and racist biases, and cases in which babies get exposed to environmental pollutants through breastmilk remind us of just how complex these issues can be. At stake is the fact that the "prevailing doctrine is that under normal conditions decisions regarding the care of children should be left to their parents or legal guardians," and that the conditions under which the state may intervene involve situations when there is "clear evidence that the action proposed would seriously endanger the child" (Kent 3). Clearly, questions of interpretation, responsibility, and control render such decisions extremely complex. The UN agencies put out a statement in 1998 recognizing that "HIV-positive mothers should be enabled to make fully informed decisions about the best way to feed their infants in their particular circumstances" ("HIV and Infant Feeding" 8). Since then, they have been even more specific in their recommendations. They clearly state that "infants should be exclusively breastfed for the first six months of life to achieve optimal growth, development and health" (WHO "HIV" 2). Addressing HIV, they specify that "given the need to reduce the risk of HIV transmission to infants while minimizing the risk of other causes of morbidity and mortality . . . when replacement feeding is acceptable, feasible, affordable, sustainable and safe, avoidance of all breastfeeding by HIV-infected mothers is recommended" (2). They emphasize: "Otherwise, exclusive breastfeeding is recommended during the first months of life" (2). Here they are careful to point out that "unnecessary use of breast-milk substitutes by mothers who do not know their HIV serostatus or who are HIV-negative should be avoided"

(2). While it seems dangerous and futile to make recommendations for all women in all circumstances, one of the few solutions agreed on by many is that when a woman, particularly an HIV-positive woman, opts not to breastfeed her child, donor milk is an important option that, according to WHO and UNICEF, "should be used" and promoted as a viable option for babies (Arnold 19). All parties could agree that promoting milk banking on a policy level would offer babies of HIV-positive women the most hope for a healthy future. Supporting milk banking, donating one's milk to existing banks, and spreading the word about the importance of donated human milk are some important steps that women can take to become involved in this important issue. And men clearly have a role in promoting banking.

THE TRUE UNBORN VICTIMS OF VIOLENCE

I noted earlier that when babies are subjected to certain contaminants through their mother's milk, they will often face greater impacts from prenatal exposures. This is important for several reasons. It suggests that even mothers who opt not to nurse will be transmitting part of their bodies' toxic load (in the case of environmental pollutants) or a possibility of their bodies' infection (in the case of HIV or other infections) to their infants. While questions about whether states should have a say in allowing certain women to breastfeed have surfaced, questions about state controls over what happens to fetuses have become a hotly contested issue. These cases have a bearing on breastfeeding because of the fact that when a mother's milk carries contaminants, her placental fluid and chord blood will often carry more (Steingraber *Having*; Gross-Loh 63). Here we see how racism, classism, and other oppressive assumptions can have a large impact on health. The court case of a group of thirty South Carolina women who were arrested, some taken from their hospital beds in handcuffs just after labor, and denied the opportunity to nurse their children is illustrative (Greenhouse). In this instance, a public hospital tested the urine of select maternity patients, nearly all of whom were black (nine out of ten) and turned evidence of their cocaine use over to the police. The possibility of positive prenatal exposure led to the arrest of the women and the curtailment of their ability to nurse. Debates around other risks to the "unborn," including those who are subjected to environmental pollutants and other prenatal violence suggests the complexity of these issues. On the one hand, there are court cases involving mothers engaging in what might be considered "risky" behavior; on the other, there are cases involving risks beyond the mother's control.

In October 2000, the Supreme Court considered the limits of the legal response to "risky" behavior by pregnant women.[21] Justice Stephen Breyer's comment that the South Carolina program "probably hurts more fetuses than it helps by deterring women from seeking prenatal care" attests to the dangers of such a precedent. The fact that one woman, Regina McKnight, was sentenced to twelve years in jail with no chance of parole because her baby was stillborn—even though she repeatedly sought drug treatment during her pregnancy and was told it was not available—indicates the problematic assumptions at stake in such a precedent. One issue involves the degree to which the baby is subjected to prenatal exposure, which often has a class and race-based component. As Julianne Malveaux comments, "If her mom is a crack-cocaine addict, the public and the media see the child's imperiled condition as criminal; if her mother is a woman who took fertility drugs and gave birth to septuplets, our culture applauds the parents and showers them with everything from free diapers to baby food for life." Malveaux points out that the difference has to do with racism and classism, and the fact that "one drug is legal and taken by choice," while "the other is illegal" (24). Referring to an article appearing in JAMA, she notes that medical personnel often remain unable to discern whether drug use is the reason for an infant's illness or death. In the JAMA issue, for example, Dr. Deborah Frank and other authors conclude: "prenatal cocaine exposure [is] no more toxic than environmental toxins or exposure to alcohol, tobacco, or marijuana" (24). Dr. Hallam Hurt, chair of the division of neonatology at the Albert Einstein Medical Center in Philadelphia, agrees, noting that "the effects of drug exposure are 'totally overshadowed by [the effects of] poverty' " (quoted in Malveaux 24). Malveaux's conclusion that "it is easier to fight drugs than to fight poverty, easier to target drug-addicted women than to admit that poverty might somehow scar their children's life chances," gets at the heart of many contemporary contamination issues. It should also be noted, in reference to this case, that when we discuss behavior that proves risky to infants, we rarely see formula use itself as posing a risk. And the cutting of the ads by the USDHHS, discussed in my introduction, makes such discussions less likely.

Discussions around the recently passed Unborn Victims of Violence Act suggest the degree to which our tendency not to see environmental effects as significant has bearing on issues that do entail environmental effects. This legislation makes it a separate crime to harm a fetus. Pro-choice advocates denounced the Unborn Victims of Violence Act as "an effort to undermine the constitutional right to abortion by recognizing the fetus as a person" (Hulse). They worried that once the definition of the "unborn" as "a member of the species of homo sapiens at any stage of development" got written into federal law, "pro-life," or "anti-abortion" advocates (the title depends on one's leaning) would use it to overturn existing laws protecting a woman's right to abortion procedures. Backers of the measure, whether or not they had abortion in mind, exhibited graphic photographs of crime victim's bodies. Now that the measure has passed, it would be healthy for us to use it in a way that would allow us to come together across political boundaries. What both sides could come to terms with is a pressing application of this legislation, as it currently stands, but one that has gone unnoticed. While many fetuses do get harmed through violent crimes committed against the mother, many also get harmed through environmental contamination that gets imparted to the fetus in utero.

Indeed, studies suggest that many women who suffer miscarriages, and whose babies are born with "birth defects" ranging from larger issues to conditions such as some forms of attention deficit disorders, face problems because their bodies contain persistent environmental pollutants (Jacobson and Jacobson 783–89; Patandin 33–41; Walkowiak et al. 1602–7). Research suggests that compounds such as mercury, dioxin, and PBDEs that get passed on in vitro negatively effect fetuses, particularly since they are in critical stages of development. But these forms of violence rarely get addressed. If punishment is at stake, we should also punish all those who support policies, practices, and legislation allowing fetuses to be subjected to such violence. We should work together to end this kind of violence against the unborn. If we want to bring pro-choice and pro-life advocates together, we must focus on those who are the true unborn victims of violence.

TWO

TOXIC DISCLOSURE

THE GROWING AWARENESS OF ENVIRONMENTALLY CONTAMINATED BREASTMILK IN THE CONTEXT OF MUCH-NEEDED BREASTFEEDING ADVOCACY

RACHEL CARSON'S LEGACY

Many people who care deeply about environmental issues know of Rachel Carson as the "mother" of contemporary environmental movements. We think of *Silent Spring*, published in 1962, as *the* seminal text that informs our thinking about the relationship between contemporary Americans and the environment in which we live. It, almost single-handedly, began to rouse us, as a nation, to the stark truth—that our approach to the world that surrounds us has begun to leave a silent, deadly scourge in its wake. Carson's nostalgic and fairy-tale opening of "a town in the heart of America" teeming with abundant plants, animals, and birds, where fields of luscious crops, fruit-laden trees, and rivers flowing with fish inhabited the land until "a strange blight crept over the area" and "things began to wither and die" speaks to our longing for a place and a time in which the profound effects of environmental degradation have not taken hold. It sparks a hope in us that we can create a world where people's decision-making always prioritizes ecosystem health and human well-being (1–2).

One of the most striking features of Carson's approach is her scientist-poet's attention, the way she speaks with such precision and awareness about the effects that human intervention has wrought, while at the same time never allowing the urgency of what she says to give readers the impression that she puts persuasiveness ahead of truth-telling. We believe her because the evenness of her tone and the wealth of details she provides suggest that she is more interested in telling the truth than in convincing us that something must be done. And yet, perhaps because of her skill in explaining complex scientific concepts in terms that all of us can understand, we realize the impact of her message and feel it nudge us toward acting more responsibly. Two examples serve to illustrate this. In discussing the use of insecticides, Carson describes several situations in which "the result of chemical spraying has been a tremendous upsurge of the very insect the spraying was intended to control, as when blackflies in Ontario became 17 times more abundant after spraying than they had been before" (252). She explains that "the broader problem . . . is the fact that our chemical attack is weakening the defenses inherent in the environment itself" (246). And then, even-handedly, she proclaims: "It is not my contention that chemical insecticides must never be used" (12). Instead, she lays out her belief that in this "era of specialists, each of whom sees his own problem and is unaware of or intolerant of the larger frame into which it fits," we fail to look for long-term consequences. "We train ecologists in our universities and even employ them in our governmental agencies but we seldom take their advice" (11–12). Carson also maintains a pragmatic eye and sees that in an "era dominated by industry, in which the right to make a dollar at whatever cost is seldom challenged," we go for the quick fix and thereby imperil our health. Faced with "false assurances" and "little tranquilizing pills of half-truth," the public must look deeper and "must decide whether it wishes to continue on the present road" (13).

Perhaps Carson's greatest contribution is her explanation of how things have changed and why urgency must prevail. As she tells it, the silencing of the spring began in the mid-1940s when "sprays, dusts, and aerosols," and other "elixirs of death" began to be used universally and without testing, "still[ing] the song of the birds and the leaping of fish in the streams . . . coat[ing] the leaves with a deadly film, and . . . lingering in soil" (7). She notes "For the first time in the history of the world, every human being is now subjected to contact with dangerous chemicals from the moment of conception until death" (15). We realize the importance of her message and its continuing importance when we think about the prevalence of the belief that "we don't really have to confront environmental problems, because—look at us—we're all fine!" She reminds us that when we fail to examine the effects on human health and ecosystem health in the aggregate, we support policies that contribute to ill health. Carson points out that polluting industries are "a child of the Second World War," and "differ sharply from the simpler insecticides of prewar days," now "enter[ing] into the most vital processes of the body and chang[ing] them in sinister and often deadly ways" (16,17).

Agents created in the laboratory for wartime use were imported for domestic use after the war. "Destroy[ing] the very enzymes whose function is to protect the body from harm, [these agents] block the oxidation processes from which the body receives its energy, they prevent the normal functioning of various organs, and they may initiate in certain cells the slow and irreversible change that leads to malignancy" (16,17). Many of these man-made chemicals, created by "ingenious laboratory manipulation of the molecules, substituting atoms, altering their arrangement," work according to the process of biomagnification, or bioaccumulation, which Carson explains with characteristic ease (16). Certain chemicals, which the body cannot break down, remain in the body's fat cells and are magnified as they are consumed in the bodies of smaller organisms by larger organisms. "For example, fields of alfalfa are dusted with DDT; meal is later prepared from the alfalfa and fed to hens; the hens lay eggs which contain DDT. Or the hay, containing residues of 7 to 8 parts per million, may be fed to cows" (22–23). Eventually, "the DDT will turn up in the milk [from these cows] in the amount of about 3 parts per million, but in butter made from this milk the concentration may run to 65 parts per million." Carson explains "Through such a process of transfer, what started out as a very small amount of DDT may end as a heavy concentration" (22–23). Quietly, without fanfare, Carson points to one consequence: "Farmers nowadays find it difficult to obtain uncontaminated fodder for their milk cows" (23). Carson explains that another "sinister feature of DDT and related compounds," many of which still persist in water, soil, and airwaves and many that continue to be released into the environment, is that those organisms that sit closer to the top of the food chain ingest toxins in their most concentrated forms.

Articulating what is perhaps the first well-publicized attention to breastmilk showing signs of environmental contamination, Carson points out that the human infant is at the very top of the food chain, subjected to the highest amounts of toxins of any organism. She explains that "the Poison may . . . be passed on from mother to offspring," so that "the breast-fed human infant is receiving small but regular additions to the toxic load of chemicals building up in his body" (23). Challenging readers to wake up to the implications of our practices, she explains that because infants ingest large amounts of nourishment in proportion to their body weight, and because the accumulation of toxins enters the body during critical stages of development, effects can be significant, debilitating, and often do not appear for years or are not recognized as resulting from environmental degradation. While many credit Carson with inciting the attention to the environment that brought about policy and attitudinal changes in the 1970s and beyond, what stands out as surprising and puzzling is that her warnings about something so fundamental as the environmental contamination of breastmilk have not caught on here. Indeed, most people have never considered the concept of breastmilk contaminated by environmental pollution or its staggering implications. Environmental groups have not, for the most part, and until very recently, seized on this poignant matter; feminists, by and large, have not embraced this issue or the lack of public support in the United States for breastfeeding accommodations in the workplace; and others we might expect to express interest, including breastfeeding advocacy groups, public health officials, governmental agencies, and the mainstream press have embraced the issue only sporadically, if at all.[1]

One of the questions this book seeks to answer is "Why is this?" Is there something about the environmental contamination of breastmilk that has kept it mostly hidden from view? Is there something about the inter-play—historically and today—between groups that one might expect to have taken an interest (for or against publicity) that has allowed this issue to be silenced or to take a back seat to other environmental or health issues that have similar or comparable risks? This chapter traces the slug-gish but growing attention to the environmental contamination of breastmilk, and argues that several factors have had an impact on the slow growth of public awareness: the interaction between changing rates of breastfeeding in the United States and episodes in which knowledge of environmental pollutants have taken hold; interactions between environ-mental groups and other pertinent groups; developments in other related campaigns, such as the movements to curb formula advertising and use and the environmental justice movement; and the sensitive and unusual nature of breastmilk on both a policy level and as a vital fluid with emotional and physiological potency.

As noted in chapter one, concerns about breastmilk contamination have surfaced at many points throughout human history, including the early portion of the twentieth century, before Carson published her master work. For example, C.B. Reed, an assistant professor of obstetrics at Northwestern University Medical School, wrote in 1908 "The removal of the nursing babe from the breast may be required for any one of a great variety of conditions which affect the mother or child or both," including "the presence in the milk of foreign substances which have been taken in by the mother, or which occur as the result of changes in the environment" (quoted in Schreiber 1114). Carson herself points to two early studies, one from 1937—"The Problem of Possible Systemic Effects from Certain Chlorinated Hydrocar-bons"—and one from 1951—"Occurrence of DDT in Human Fat and Milk" (Drinker et al.; Lang). But awareness among scientists does not translate into more general public awareness, policy shifts, or changes in behavior, which must take place for us to put an end to polluting activities.

If we consider instances in which the environmental contamination of breastmilk has appeared in major U.S. newspapers during the period 1960–1980 as one indicator of awareness in those early years after Carson published *Silent Spring*, the sparse number implies that, despite some sig-nificant early mention, the issue did not achieve public prominence in this period. In addition, the specifics of this attention—for example, the deci-sion by the Environmental Defense Fund to feature this issue prominently and then to back away from it—suggests that such attention is character-ized not by a growing urgency as more people gain awareness, but by a complexity involving the history and nature of how breastfeeding has been framed in the United States. During the period, papers in the northeast region offered seven mentions of breastmilk, with twenty-nine in the southeast and none in the twenty-year period in western and midwestern states.[2] After Carson, the first public mention of environmentally contaminated breastmilk that I have been able to locate appears, after seven year's silence, in a 1969 New York Times article that notes findings by a Swedish doctor that average daily concentrations of dichloro-diphenyl-trichloroethane (DDT) in the body fat of people throughout the world means that breastfed infants "ingest about twice the amount of DDT recommended as the maximum daily intake by the World Health Organization" (6 May 1969). Then, suggesting that the issue might achieve the prominence it deserves, in 1970, the Environmental Defense Fund (EDF) used toxic breastmilk as the call to action for its first membership campaign ("Where It All Began"). The focus made sense for EDF, after having been founded in 1967 around the fight to ban DDT, which it accomplished in 1972. The EDF's campaign featured a headline in the New York Times—"Is Mother's Milk Fit for Human Consumption?"—and came out of the recognition that DDT levels in human milk had risen to seven times that permitted for milk sold in stores (Silbergeld).

A later recounting of EDF's first membership campaign hints at part of the reason for the long lag time between Carson's warning and subsequent attention. Noting that "her book was a best seller, but the government did nothing to prevent DDT and a myriad of other chemicals from contaminating the environment," it implicates the reality: that most Americans, on first encountering evidence of environmental despoilation, expected that "the government" would simply "take care of it" ("Where It All Began"). Celene Krauss's "Women and Toxic Waste Protests: Race, Class and Gender as Resources of Resistance," which compares interest in environmental issues by three groups—white, African American, and Native American women, and finds white Americans confront environmental issues trusting that the government will simply take care of environmental problems—adds weight to this reading. In today's more cynical, post-Watergate world, we might find such trust to be naïve, but it defined public attitudes in the past. In addition, we must remember that agencies and legislation to confront environmental problems had not yet been established in Carson's day; the Environmental Protection Agency (EPA), for instance, was not created until 1970.[3] The wording of the next visible mention of toxic breastmilk, in the New York Times in 1976, points to the general lack of awareness on the part of many citizens at this time to environmental toxicity. It reports that "a chemical called PCBs has been found in mother's milk," wording unthinkable today when the mention of polychlorinated biphenyls (PCBs) is familiar to most readers (24 September 1976).

THE IMPACT OF CHANGING RATES OF
BREASTFEEDING IN THE UNITED STATES ON INTEREST
IN ENVIRONMENTAL POLLUTANTS IN BREASTMILK

Another, at least partial, explanation for the lack of interest in the issue in
the period immediately following *Silent Spring* seems to be that during this
time people were not just not interested in the environmental contamination of breastmilk, they were not interested in breastmilk at all—or at least
most people were not. We must bear in mind that whereas 90 percent of
babies had been breastfed in the United States in 1920, by 1946 only 38
percent of mothers nursed after their discharge from the hospital, with only
5.5 percent continuing to do so at five or six months (Steingraber *Having*;
Carter 3–4). The low point, in fact, coincides with the publication of Carson's
warning, with 20 percent nursing at birth in 1959, and 18 percent in 1966,
with very few continuing for several month's duration. By 1971, the number
had risen somewhat, to 25 percent in the initial period after birth. And, by
1981, the high point since the 1930s, 58 percent of women were nursing
initially, with 24 percent continuing through a few months. When we consider these numbers, we must recognize that the low rates of breastfeeding in
the United States have a complex history, including the growing
medicalization of childbirth and childrearing, the powerful lobbying by formula manufacturers, and the nature of gender issues. This history is important to bear in mind today, when the percentage of women nursing has fallen
since the high point in 1981. Currently, only 20 percent of mothers are still
nursing when their babies reach five months and fewer than 5 percent are
still breastfeeding when their babies reach one year (Steingraber *Having*;
Wolf 2007; Ruowei et al.).[4] As we will see, calling attention to the environmental contamination of breastmilk holds risks in an environment where
breastfeeding rates are low, putting pressure on breastfeeding advocates who
sometimes lobby those who might bring attention to the issue.

This interplay between interest in the environmental contamination
of breastmilk and interest in breastfeeding itself plays out when we consider
Europe. In contrast to the United States, more people in Europe regard
breastfeeding as important (Platypusmedia.com; WHO "Global"). With less
pressure to portray breastfeeding in a completely "untarnished" light, environmentalists and breastfeeding advocates there can more freely address toxicity. Advocacy groups, such as Women in Europe for a Common Future
(WECF), have taken an active role in working to eliminate the production
of PVCs; they note in a letter voicing their protest that "our offspring will
sooner or later experience the health effects from this contamination" (Letters Women 3).

If we compare attention to breastmilk toxicity in the United States
during the periods in which rates of nursing attained their high point—the
early 1980s—to periods when fewer women nursed—the late 1950s and early

1960s—patterns do not fully play out according to expectations. While it makes some sense that attention to environmentally contaminated breastmilk might be sparse right when Carson was publishing, given that this was the era in which more women relied on formula than at any other period, one might expect the rate to pick up, and it would follow that the period in which the largest proportion of women chose to nurse would have constituted the heyday in concern for environmental contamination. But while some of the major newspapers and environmental groups devoted some concern for toxic breastmilk at that time, such coverage was limited and got overshadowed by other issues.

For example, a September 25, 1980, *New York Times* article titled "Toxic Risks to Babies Underlined" reports on a conference sponsored by McGraw-Hill, publishers of a recently released book *At Highest Risk: Environmental Hazards to Young and Unborn Children*. The story notes that, at the conference, the theme that "women must mobilize to fight the toxic hazards facing their young and unborn children" "was repeated again and again by the speakers." But what we find is that reporting on two other related issues, or reporting on that issue, albeit a very localized version, took center stage. For example, in 1981, four articles (appearing December 6, May 24, May 21, and May 19) in the *New York Times* reported on the heated controversy being waged around the marketing of baby formula and the vote by the World Health Assembly to impose a code to restrict such marketing, particularly in developing countries (Brozan; "Courageous American"; "Diplomacy"; Solomon). Indeed, this issue assumed such prominence then that when you mention "infant feeding" to anyone of a news-consuming age during that period, their first thought is "the Nestlé boycott." In that same period, reports on the contamination of cow's and human milk in Hawaii because of improper spraying of the pesticide heptachlor on pineapples allowed mainland readers to see breastmilk contamination as a localized issue, something occurring far away with an easily controllable, definite end. The title of the piece—"Contaminated Milk Problem in Hawaii Nears End"—offers a tidiness to the issue, so that readers (except those residing in Hawaii) can remain distant from the problem, imagining that it does not affect them and that, even if it did, it could easily be remedied (23 May 1982). During this period, one can witness a tendency for studies to be geographically narrow, and for news coverage to reflect that.

Similarly, another September 25, 1980, article—this one in the *New York Times's* "Science Watch" section — reports that "a study of milk from more than 1,000 nursing mothers in Michigan found the environmental contaminant polychlorinated biphenyls (PCBs) in every sample," again allowing readers to regard the issue as localized and one relegated to the industrial Midwest and easily remediated—as in the Hawaii story—not as an issue that affects us all, wherever we live (Klemesrud and Dooley). Extensive coverage in the papers of another heptachlor debacle, five years later, when tainted feed got distributed to cattle in Arkansas, Missouri, and Oklahoma, reveals a similar rendering as a local issue, and as one that, once confronted, could be quickly resolved.[5] The headline in an April 11 *Arkansas Democrat-Gazette* article, "Arkansas Health Department, CDC Offer Good News for Milk Drinkers," in which readers are told that an "all clear" had been issued for the consumption of dairy products, and an April 29 article beginning "Pesticide levels in lactating women in Arkansas are about the same as those for other people across the nation," suggest that the issue is limited to this one area, and easily containable ("Breast Milk Tests Show"). The body of the latter article, however, in noting "a heptachlor tolerance level for breast milk has not been set," indicates the degree to which assurances of safety have to be taken with caution, and reminds us of the arbitrary nature of "tolerance levels" on which assurances of "safety" rest. That a federal Environmental Protection Agency report published in 1980 stated that a five year study in the United States "indicated heptachlor epoxide"—the same carcinogen as in the Hawaii and tristate incidences, and one that Carson discusses at length—"can be found in the breastmilk of over 90% of the U.S. population," and yet was not covered in any major newspaper or by any environmental group at the time, demonstrates that at this opportune moment the issue was not garnering the attention it could have, based on studies that were being conducted ("Breast Milk Tests Show"). Similarly, no press or environmental groups took notice of a 1981 U.S. government-commissioned task force on breastfeeding and human lactation, which found that breastfeeding reduces mortality and morbidity in all settings, coverage that could have built on the momentum in breastfeeding rates occurring at the time (Schreiber, note 79).

BURGEONING ENVIRONMENTALIST
ATTENTION AND PULLING BACK

Earlier, in 1977, another governmental task force, a senate subcommittee that held a public hearing, did gain coverage, and also involved the prominent presence of an environmental group—again the Environmental Defense Fund (Brown). Interestingly, coverage of the hearings in the *Washington Post* considered contamination of both human milk and cow's milk, and even linked the two, a rare occurrence since those seeking to play down the environmental contamination of human milk often quickly point to the contamination of cow's milk or formula as a counterpoint, not acknowledging that if one is contaminated, so, most likely, will the other be (Brown). The story began: "Scientists told a Senate subcommittee yesterday that it is difficult, if not impossible, to find uncontaminated milk for newborn infants," and goes on to note that while "human breast milk increasingly contains pesticide residues and other chemical contaminants that can cause cancer and other diseases," "harmful lead deposits often are found in formula." Stating that "cow's milk" is "itself a major source of the contaminants found in human breast milk," Dr. William B. Weil, Jr., professor and chairman of the Department of Human Development at Michigan State University, speaks in the vein of Rachel Carson: "As we have contaminated our environment, we have contaminated our bodies and in doing that we have contaminated human breast milk." While the committee's chair, Senator Edward M. Kennedy, notes that "the purpose of the hearings is not to discourage breastfeeding," EDF representatives, who also presented a report at the hearings, observed that "pesticide residues, ingested through the consumption of vegetables and meats, are usually present in human milk more frequently and in higher levels than in cow's milk," and seemed to be discouraging breastfeeding for the most exposed women.

The EDF's report, "Birthright Denied: The Risks and Benefits of Breast-Feeding," discussed in the article, at the hearing, and in other news items at the time, states: "While under normal circumstances EDF would support a parental decision to breast feed a child, these are not normal times so this recommendation will have to be tempered" (Brown). Although Weil made a similar recommendation, responding, "Should I breast feed my baby under these circumstances?" "I have no easy answer. . . . I really don't even have a good complicated answer," the EDF seems to have become associated with the position that questions the benefits of breastfeeding in certain cases. For example, on July 7 of that same year, the *Washington Post* ran another article in which a couple, both announced as working for the EDF, again questioned the advice that breastfeeding should still be recommended across the board, despite heavy contaminants (Burros). Stephanie Harris, the wife, notes: "EDF cannot unequivocally recommend breastfeeding to every woman because of the chemical residues that find their way into breast milk."

Close observation of the EDF's handling of the issue reveals that the group seemed to back away from breastmilk contamination for a period, and that, with time, it "tempered" its early response. And this also seems to be true for the few environmental groups that have weighed in on the issue. In the few instances of early attention, if a group or an individual questioned the advice that breastfeeding should be recommended despite contaminants, they either backed off from that questioning in time or avoided the issue altogether. For example, while the EDF sued the EPA under the Toxic Substances and Control Act to take concerted action to reduce exposure in 1985, in 1987, when it held a news conference to release the findings of an important new study on breastmilk contamination (the Schector study), it supported breastfeeding wholesale (Shabecoff; Silbergeld; noted in Harrison 48). The EDF stated that they "would not recommend that mothers stop nursing babies, even though there [is] much less dioxin in cow's milk and commercial baby formulas" since the "benefits of nursing probably exceed the risks" (Shabecoff). The EDF suggests that it has rethought its position either because of new findings, because it perceived itself to be too closely associated with the position that breastfeeding in the face of environmental contamination is not always preferable, which has become linked, by some, with antibreastfeeding sentiment, or because it was responding to being accused of fear-mongering for the sake of furthering environmentalist causes.

Katheryn Harrison, in an article published in *Policy Sciences* that focuses exclusively on the dioxin contamination of breastmilk, devotes some attention to the questions of whether environmental groups have indeed engaged in fear-mongering for the sake of environmental activism, whether they have reacted in anticipation of such claims, or whether it makes sense to draw a link between attention to the issue and antibreastfeeding sentiment. She points out that while "Both the Canadian and the U.S. governments have determined that breast-fed infants are among the population most exposed to dioxins, receiving levels of exposure orders of magnitude above those considered acceptable . . . the issue has received less media and governmental attention than other environmental issues believed to present comparable or lower health risks" (35). Arguing that "North American environmentalists have consciously chosen not to press the dramatic issue of breastmilk contamination out of concern that mothers would discontinue breastfeeding, as well as personal anxiety about an issue that fundamentally challenges conceptions of our own bodies and our relationship to our children," Harrison narrates a complex story of inattention to dioxin toxicity. Asserting that this is an issue of "self-restraint" on the part of environmentalists, that "their self-restraint challenges the depiction by some authors of environmental groups as eager to capitalize on any opportunity to provoke public concern and outrage to advance their agenda," Harrison gets at an important part of the story (35). Here, environmentalists, wary of inciting women to refrain from nursing and "personally" uncomfortable with the issue, opt to remain silent. She concludes: "The absence of significant press coverage . . . suggests that environmentalists have not attempted to publicize this particular issue, whether as a concern in and of itself or as a way to mobilize public concern for persistent toxic substances more generally" (44). Although evidence suggests the viability of this narration of the situation, other evidence maintains that such silence may also be linked to a reluctance to leave themselves open to further characterization as being responsible for feeding "hysteria" as well as the workings of a subtle and perhaps unspoken "blackmail."

Several occurrences do suggest that this makes sense. While some groups—in particular the EDF, the Natural Resources Defense Council, Greenpeace, and the Sierra Club—have addressed toxic fish and pesticides impacting children's health, environmentalists have, for the most part, stayed away, despite the poignancy of toxic breastmilk. Charges of hysteria also abound, as we can see in two examples. In *Ecoscam*, PBS Producer Ronald Bailey points to "three seminal" environmental texts—including *Silent Spring*—and claims that these "gloomy books . . . have dramatically skewed public policy for the past two and a half decades, slowing economic growth and unnecessarily increasing human misery" (1). Arguing that "a cadre of professional 'apocalypse abusers' frightens the public with lurid scenarios of a devastated earth, overrun by starving hordes of humanity, raped of its precious nonrenewable resources, poisoned by pesticides, pollution, and genetically engineered plagues, and baked by greenhouse warming," Bailey's mid-1990s text gets echoed six years later in Peter Huber's "rhetoric of hysteria." Claiming "the models can link any human activity, however small, to any environmental consequence, however large," and announcing that "it is just a matter of tracing out small effects through space and time, down the rivers, up the food chains, and into the roots, the egg shells, or the fatty tissue of the breast," Huber's discourse suggests that when environmentalists exhibit a reluctance to speak out, part of this may be due to the perception that embracing such a poignant issue as a cause célèbre might backfire (xiv).

ENVIRONMENTALISTS' RELUCTANCE

Environmentalists' relative silence on breastmilk toxicity involves many factors, including a reluctance to risk promoting formula-use or appearing to do so, as well as an unwillingness to find themselves accused of being antinursing. Indeed, a series of interactions between environmental groups and breastfeeding advocacy groups suggest that such concerns have from time to time been operative. To add weight to this reading, we can see that more than one environmental group has addressed the environmental impacts of formula production and use at various points, and that they have visibly promoted alliances with breastfeeding advocacy groups at other points, thus implying their desire to publicly acknowledge that they do, indeed, support breastfeeding—especially when they broach the subject of breastmilk contamination. An article in the September/October 2001 issue of the *Sierran*, titled "Breast-feeding and the Environment," notes that "one thing about breast-feeding that is less emphasized is the environmental benefits" (DeBarthe). The article points out that breastfeeding "requires no packaging . . .[that] Trees don't need to be cut down to make cardboard and aluminum doesn't need to be mined" and that "every 3 million babies that are bottle-fed use 450 million cans of formula requiring 70,000 tons of metal." And it suggests that Sierra has an interest in aligning itself with breastfeeding advocacy groups, rather than in giving even the impression that in pointing to the environmental contamination of breastmilk (as it had done at one point, but not in this article), it might be promoting formula-use. The article ends with the advice that mothers "contact your local La Leche League for more information."

In addition, Greenpeace, which has weighed in on the issue of breastmilk contamination or, more often, related issues such as fish contamination, has seemed to respond to accusations that it might be viewed as promoting formula-use if it raised attention to breastmilk toxicity. Harrison tells of a

situation in 1997 when Greenpeace USA "held a press conference at which
an association between PVC and dioxin in breast milk was expected to arise"
(48). Seemingly in anticipation of characterizations of their approach as
antibreastfeeding, the organization went out of its way to invite representatives of La Leche League (48). Harrison notes: "Whatever efforts environmental activists may have made to focus public attention on the issue of
dioxin in breast milk as a symbol of more widespread risks from persistent
toxic substances were rebuffed by more dominant messages: That 'breastfeeding
is good' and that anyone who might hint otherwise is at best irresponsible
and at worst malicious" (50). Indeed, she describes a situation in which
Greenpeace Canada inadvertently raised the issue of dioxin in breastmilk in
its dioxin in food initiative. Looking at an internal memorandum in which
the campaigner responsible for the initiative became encouraged by the response and developed a proposal for an ongoing campaign focused on the
risks of dioxin to children "as the most defenseless and vulnerable members
of our society," she describes how the campaigner, who "thought this was a
great environmental issue, because of that angle," finds his colleagues "dis-
playing uncharacteristic restraint for an organization that has made its name
with publicity-seeking protests" (51–52).

     While the EDF used the environmental contamination of breastmilk as
the subject of its first membership campaign in 1970, addressed the issue in
1977, petitioned the EPA in 1988 to take action to reduce exposures to
dioxins and furons, and held a news conference in 1987 around the release
of the Schector study, its 1988 coverage about dioxins and furons in human
milk seems to constitute the beginning of a more permanent phasing out of
the group's attention to breastsmilk toxicity. Greenpeace, on the other hand,
has focused more on toxics issues in general; it initiated a campaign in 1988,
called "Testing for Toxins," and it held the "Great Louisiana Toxics March,"
a march down the area dubbed "cancer alley" because of the heavy concentrations of industry and cancer. And in October 1989, the group protested
chlorine bleach, and in 1990 it began work, in Texas City, Texas, on the
Sanders Housing Project situated next to an insecticide plant. In 1995, it
focused on the Global Treaty to Reduce Toxics and worked to keep the
largest proposed PVC company, owned by the Shintech Corporation, out of
Convert, Louisiana. In 2000, when UN officials announced that 122 countries had agreed on a treaty banning twelve highly toxic chemicals labeled
the "dirty dozen," Greenpeace called this the "beginning of the end of toxic
pollution" (December 11, 2000, The Record, Bergen County, NJ). But in all
these cases, the focus on toxics did not include attention to breastmilk.
Finally, in 2002, Greenpeace weighed in when it posted a story on its website
on persistent organochlorine pollutants (POPs) accumulating in the breastmilk
of indigenous peoples. If this recent attention by Greenpeace can be taken
as an indicator of where we are going, it offers hope that the kinds of coverage that have been lacking are beginning to surface.

Similarly, in 1976, the Sierra Club addressed issues related to toxic breastmilk in its attention to the Fox River people, Southeast Asians who could not read warnings published in English telling people to refrain from eating fish caught in contaminated lakes. As noted in chapter four, its coverage, like that of some other groups, seems to have been instigated in response to burgeoning attention to environmental justice issues. As one might expect, the National Wildlife Federation's attention to human breastmilk came out of its focus on toxic fish; the group took up the issue with a 1990 article "Lake Michigan Sport Fish: Should You Eat Your Catch?" that mentions breastfeeding mothers as a population vulnerable to the effects of eating Lake Michigan fish. However, the group seems to have dropped the issue thereafter.

CHILDREN'S HEALTH CAMPAIGNS: THE POWER OF THE INTERNET

In 1991, we saw a new phenomenon in attention to breastmilk toxicity by environmental groups with the establishment of CCEF, the Colette Chuda Environmental Fund, by the parents and friends of Colette Chuda, a four year old who died of cancer thought to be associated with environmental causes. This group and others like it emerged as a response to increasing rates of childhood diseases and childhood cancer and the growing suspicion that such disorders can be linked to environmental causes. The Chudas tell of how they discovered a study published in the *American Journal of Epidemiology* that "revealed a link between parental pesticide use" and the development of tumors, and suggested that "the effects of pesticides could be mediated by mutations in germ cells, by exposure of the fetus in utero, or by exposure after birth from residues present in breast milk, in foods, in the home, or in the surrounding environment" ("Our Story" 1). In 1991, they launched CCEF to support research on the risks to children from environmental toxins. Building on the work of feminists who had earlier fought hard to bring attention to the reality that scientific studies typically ignore women's bodies so that effects on women frequently get ignored, they strove to bring attention to the fact that "virtually all of our environmental protection standards in the United States are based on research that measures the potential effects of carcinogens on a 155 pound adult male."

In 1994, the group released its first systematic analysis of children's exposure to carcinogens in their home and school environments, with a study *Handle with Care: Children and Environmental Carcinogens* that was researched by the Natural Resources Defense Council (2). While continuing to develop scientific research, the group, currently called CHEC, the Children's Health Environmental Coalition, works to "mobilize parents and caregivers, environmental groups, the scientific community and media to work together to help make the necessary changes in national policies to protect our children." Filling a gap left by others, this group seeks to educate people and to "share some of the practical strategies and alternative products" that could help minimize environmental health risks to children. CHEC continues to work on breastmilk issues. On November 11, 2002, for example, it reported that "an international panel of 30 experts met at the Penn State College of Medicine . . . to develop a plan for a nationwide effort to discover what, if any, environmental chemicals transfer from mothers to their babies through human milk" ("Researchers Look" 1). Dr. Lawrence Gartner, chair of the executive committee of the Section on Breastfeeding of the American Academy of Pediatrics, notes that "human milk feeding is of such critical importance to infant, child and even adult health that every effort must be made to assure its safety and purity," while he assures us that, to date, only acute massive intakes of contaminants by breastfeeding mothers have been associated with infant disease (1–2). Ultimately, the group's goal is to work to "secure funding for pilot studies" and "get the facts about the presence or absence of environmental chemicals in human milk" (2).

In 1992, a related group, the Children's Environmental Health Network (CEHN), formed as a national organization. By this time "mainstream" environmental groups seemed to have either moved away from the issue of toxic breastmilk, as did the EDF, or to have taken up the issue—but from an environmental justice angle—as with the Sierra Club. CEHN's inception represents a phase in which groups focusing specifically on children's health and the environment took up toxic breastmilk as one of several issues linking children's health and the environment. Recognizing that "children are not little adults," that they have rapidly developing systems, breathe more quickly, have higher metabolic rates, and that they eat and breathe more per body mass weight than adults and are extremely vulnerable, CEHN strives to provide education and influence policy that assure the rights of all children to a healthy life (Witherspoon). Examples of the kinds of work the group has done include its effort to bring about the 1997 Executive Order on Children's Environmental Health and Safety, its input on the creation of the Office of Children's Health at the Environmental Protection Agency (EPA), its sponsorship of the 1996 Food Quality Protection Act (FQPA), the most protective legislation for promoting the setting of allowable thresholds at levels safer for children, and its support, in 1998, of the establishment of Pediatric Research Center and Pediatric Environmental Health Specialty Units. One talk that I attended, presented by public educator Nsedu Witherspoon, discussed the group's educational campaigns and its focus on minority populations and the "realities of racial disparities in environments and health"; the group continues to bring articulate, informative discussions to people throughout the United States.

My intent is not to blame environmentalists or to find fault, any more than it will be to blame or find fault with feminists later in the discussion, for not devoting more attention to this issue. Both environmentalists and feminists have done invaluable work for years helping to raise awareness, shape public policy, and forge creative strategies that have made this country a more tolerant, less oppressive, more responsible, and healthier place to live. Certainly, the relentless work performed by countless environmentalists has helped ensure that breastmilk today is not more subject to environmental contaminants. We have environmentalists to thank for innumerable shifts in policy and practices that have rendered our country safer. My investigation into shifting attention and inattention to breastmilk toxicity suggests that, left to themselves, environmentalists (and feminists) might well have embraced the environmental contamination of breastmilk as a key issue long ago. But we inhabit a complex world, where issues get politicized and situations get affected by complex webs of cascading events.

IMPACTS OF THE BREASTFEEDING ADVOCACY COMMUNITY

Breastfeeding advocacy groups play an important role in this narrative in three ways. First, these groups have been important in bringing attention to breastmilk as a vital part of parenting and as crucial to infant well-being. Without advocacy groups, breastfeeding rates would almost certainly remain low. Again, for breastmilk toxicity to be taken seriously, people must care about breastmilk and see it as a substance worth protecting. Second, these groups have acted as key players in the campaign involving another form of infant food contamination—the campaign against formula manufacturers and the kinds of problems associated with formula-use. The rise of corporate, industrialized formula production, along with the colonizing approaches associated with the marketing of infant formula worldwide, become key elements in this narrative. Third, these groups occupy an important position in that their actions have an impact on whether and how environmental groups have raised the issue of breastmilk contamination. Sometimes some advocacy groups have been so invested in promoting breastfeeding that, fearing that the subject of breastmilk contamination would frighten women away from nursing, they have acted to suppress the issue, depriving the public of knowledge that could help parents and others confront the world in which we live.

Many branches of the breastfeeding advocacy community, a conglomerate of voices and movements, have done a wonderful job of focusing much-needed attention on the uncontested benefits of breastmilk in a range of settings. Responsible for campaigns to educate people and initiate policy shifts to mandate accommodations for nursing moms working outside the home as well as a wide range of other programs, these groups—including La Leche League International (LLIL), International Breast Feeding Action Network (IBFAN), Infant Formula Action Coalition (Infact), Baby Milk Action, and a host of others—provide an invaluable service. They have worked hard to bring people to understand the far-reaching effects of breastmilk and to help promote a breastfeeding culture in the United States so that women can now nurse in public without fear of harassing comments or prosecution.

These groups have also worked and continue to work on much less-publicized campaigns; for example, Baby Milk Action's use of sophisticated materials to assist teachers in raising student awareness of insidious corporate messages—called *Seeing Through the Spin*—is just one example of the breadth of work these groups have undertaken to promote critical thinking, responsible practices, and healthy people (Richards). Had the trends toward formula-use that began to accelerate in the 1930s been allowed to continue unchecked, breastfeeding—which clearly is a learned skill—might well have disappeared. Even still, it does not occur to many women to think of breastfeeding as an option, while many others regard it as an option comparable to using formula, and so, accustomed to cultural habits that promote formula-use, they choose formula. Having begun to breastfeed my first child after I had been researching and writing this book for quite some time, I can attest, from firsthand experience, that one does not simply "know" how to nurse. If a woman expects nursing to be easy and come "naturally," and then finds that her nipples are sore, cracked, and bleeding, her breasts engorged, and does not know how to deal with the problem or know that the problem will soon, most likely, end, she may not continue nursing.[6] Or if she begins to fear that her child is not gaining adequate nutrition, she may opt for formula, of which she most likely has free samples in her goody-bag of freebies from the baby food industry. While breastfeeding advocacy has come a long way, much work still remains to be done.

DEVELOPMENTS IN RELATED CAMPAIGNS: MOVEMENTS TO CURB FORMULA ADVERTISING AND USE AND TO PROTECT INFANT HEALTH

Part of this work involves bringing attention to the colonizing gestures that have lead to the exporting of formula, particularly to developing nations, as well as those actions that have brought about the spreading of the industrialized world's ways—including our short-sighted approaches to the "natural" world, and our exporting of wastes and polluting industries abroad that have all impacted both breastfeeding rates and the purity of breastmilk the world over. Any discussion about "toxic" infant food must include infant formula as well as debates about the marketing of breastmilk substitutes by the formula-producing industries. Indeed, focus on the issue in the *New York Times* in 1981, which depicts the heated tone of the debate, reminds us of the degree to which, to many, mention of "toxic infant food" means infant mortality and morbidity produced by the inappropriate marketing of formulas, especially in contexts where women are more vulnerable. Stephen Solomon, contributing editor to the *Science Digest*, writes in "The Controversy Over Infant Formula" that the nerve center of the protest became the Interfaith Center on Corporate Responsibility, but that "the words that each side used against the other are not the kind usually heard among the pews on Sunday mornings." He continues: "So vituperative is the rhetoric that it is easy to imagine both sides settling the issue in a back alley after dark." Solomon reminds readers that "at the center of the increasingly bitter conflict are babies, millions of babies with the shriveled limbs and the distended bellies that signal kwashiorkor, the Ghanian term for malnutrition that has become part of the medical literature." At the time this article was published, UNICEF established that eleven million infants died each year in developing countries before their first birthday—and noted that some estimates "blame reliance on infant formula for about a million of these deaths."

At this point, I want simply to provide an overview of a few of the most significant events in the history of the "campaign to protect infant health" in order to suggest some of the contours of the long battle that advocates have waged, sometimes with great force, against the formula industry. In 1939, in one of the first and most contentious campaigns, Cicely Williams presented a talk on infant deaths from formula; calling her speech "Milk and Murder," she claimed that "misguided propaganda on infant feeding should be punished as the most criminal form of sedition," and argued that "those deaths should be regarded as murder" (Richards 108). While those carrying forth the campaign, for many years, were individuals acting alone or in small groups, nationwide groups eventually formed and, by the 1970s, international agencies got involved. For example, in 1970, the UN Protein-Calorie Advisory Group (PAG) "raise[d] concern about industry practices" and set out to combat the issue on a international level (108). The year 1974 marks the date in which War on Want published "The Baby Killer," a contentious report pointing to infant malnutrition and blaming the promotion of artificial feeding in the third world; that same year Nestlé began to fight back, suing for libel. While a court found the Third World Action Group, who translated "The Baby Killer," guilty of libel for the title only, it also ordered Nestlé to change its marketing practices (109).

Soon after, the Infant Formula Action Coalition (INFACT) launched the Nestlé boycott in the United States in 1977, which spread to much of the rest of the world in the following years. Things began to look hopeful when Bristol-Myers settled a lawsuit in which it agreed to end all direct consumer advertising of baby milks and halt its practice of using company representatives as mothercraft nurses in its attempt to get mothers to use its products. But this curtailment on consumer advertising did not extend to the United States and has frequently been abandoned in other parts of the world. That is also the year in which Senator Edward Kennedy held US Senate hearings on the inappropriate marketing of infant food in developing countries, and in which the public became aware of such marketing as a huge problem. The next year, WHO/United Nations Children's Fund (UNICEF) hosted an international meeting on the feeding of infants and young children that called for the development of an international code of marketing. At that meeting, Baby Milk Action founded the International Baby Food Action Network (IBFAN), which started to recruit other groups.

Finally, in 1981, the thirty-fourth World Health Assembly (WHA) adopted resolution WHA$_{34.22}$ that includes the International Code of Marketing of Breast-Milk Substitutes as a "minimum requirement" to be adopted "in its entirety"; 118 nations voted in favor, with only the United States voting against (110). By 1984, under pressure, Nestlé agreed to implement the International Code in developing countries, and advocacy groups suspended their boycott. Unfortunately, reports of violations of the code continue to be noted. In 1986, the thirty-ninth World Health Assembly took action and adopted a resolution banning free and subsidized supplies of breastmilk substitutes, but the next year IBFAN monitoring "reveal[ed] companies flooding health facilities with free and low-cost supplies" of substitutes that encourage women to desist breastfeeding. To put these actions into perspective, we must recognize that once a woman's milk begins to dry up, reestablishing her flow can prove challenging, if not impossible (111). The transition happens very quickly, so that when a woman supplements a few times using formula, her milk will begin to be less and less abundant.

As the 1980s progressed, efforts continued in the campaign. In 1988, the U.S. IBFAN group launched another boycott against both Nestlé and Wyeth/AHP (American Home Products) in the Untied States, and the World Health Assembly lamented "continuing decreasing breast feeding trends in many countries." The following year, assuming a new strategy, the United Nations adopted the Convention on the Rights of the Child, focusing on breastmilk as a "right" while the next year the World Health Assembly found that "free or low cost supplies continue to be available to hospitals and maternities" (111–12). In an important move, UNICEF and WHO, in 1991, launched the Baby Friendly Hospital Initiative, an effort to end free supplies of infant formula in hospitals, and to move hospitals toward practices that promote nursing. In its *State of the World's Children* published that year, UNICEF said that reversing the decline in breastfeeding could save 1.5 million lives every year (112). As companies continued to market in what many see to be "unethical" ways, IBFAN published *Breaking the Rules* in 1994, a result of monitoring in sixty-four countries; today it continues to keep track of rule breaking. IBFAN itself underwent monitoring in 1997 to ascertain whether its process of overseeing through the years had been accurate; the resulting report, called "Cracking the Code," found that the International Code continues to be broken in a "systemic rather than on-off manner" (113).[8]

Today, efforts continue in the work to check the marketing of infant formula. It is important to reiterate that while the International Labor Organization's resolution against direct marketing of infant foods has been passed in numerous countries, the United States remains one of the few nations that still allows direct marketing. Anyone who is pregnant, has had a baby recently, or has adopted one, can attest that the marketing of formula here is relentless. As I write this, I have received three 12.9 oz. cans of powdered Enfamil, three 3.6 oz. cans and two 12.9 oz. cans of Similac, along with two very nice diaper bags complete with changing pad and yellow zipper pouch, and one CD of classical music from Enfamil, which tells me that Enfamil is "dedicated to each baby's healthy growth and development." Most of these items have come in the mail, while one pouch, a larger can, and the CD were from a pediatrician's office where my husband was doing his third-year medical school rotation.

If we return to Solomon, who covered the issue of attempts to regulate direct marketing in the 1981 *New York Times* article mentioned earlier, we find the reminder that formula, when used appropriately, can be a lifesaver. But, the article reports, it can be an acceptable alternative "to breastmilk only under certain conditions: when the mother can afford to buy sufficient quantities, when she has access to refrigeration, clean water, and adequate sanitation, and when she can understand the directions well enough to mix the formula properly." Solomon reminds readers: "Those critical of the formula producers point out that most women in developing countries are not in the position to use formula safely," so that many, many babies become ill, malnourished, or die as a result. He also notes that many recognize that formula-use can have negative effects in the United States and other developed countries. Indeed, many claim that aggressive tactics by formula companies—from dressing salespersons as doctors and nurses, a practice used early on, to giving away samples so that women lose their supply of breastmilk and may not be able to reestablish it—have "contributed to a net shift away from breastmilk, the safest and most nutritious food for infants."

I want to return to the significance of the fact that the period in which the rates of breastfeeding were the highest in the United States (the early 1980s) was also the period in which the Nestlé boycott and the debate over the marketing of infant formula in "developing" nations became the most pronounced. With more women nursing than at any other period since before the reign of "scientific motherhood," things ripened, in the United States, for attention to the environmental contamination of breastmilk. In addition, also at this time, scientists conducted several studies documenting breastmilk toxicity—thus setting the stage for wide attention to the issue of environmental contamination and for people to become outraged over that issue. But, instead, this other field of developments began to take center stage. The situation became exacerbated when some prompted the perception that if you raise the issue of breastmilk contamination, you are also, by extension, promoting formula-use—and this had become unacceptable. That more and more U.S. women began to breastfeed and that the media began to focus more and more attention on nursing makes the curtailment of attention even more poignant. To give just a few examples, a March 31, 1981, "Science Watch" article in the *New York Times* reported on comparative rates of breastfeeding and noted that it was "on the rise," while a September 21, 1980, article described an area mall that had created a private nursing room, thus bringing well-deserved attention to the act of nursing that women faced on a daily basis (Hanley "New Jersey"). In August 1985, the *Washington Post* ran a piece noting that the American Academy of Pediatrics recommended breastfeeding for at least six months, thus holding out physician approval as an added incentive to breastfeed ("Prevention"). And in April 1986, the *New York Times* cited a decline in breastfeeding as partly responsible for a "vicious cycle of infection and malnutrition in children," alerting moms to the documented benefits for children (Altman "Doctor's").

This could have been the perfect moment for attention to the environmental contamination of breastmilk to have taken hold. And indeed, several important studies were released. As already noted, a federal Environmental Protection Agency (EPA) report published in 1980 noted that a five-year study in the United States indicated that "heptacholor epoxide can be found in over 90% of the US population" ("Breast Milk Tests"). But none of the major newspapers or environmental groups commented on this finding. In 1985, one of the most important studies showing the reality of breastmilk contamination was published—Rogan's "Study of Human Lactation for Effects of Environmental Contaminants: The North Carolina Breast Milk and Formula Project and Some Other Ideas"—but it did not get coverage in the major newspapers or by environmental groups. Instead, coverage from this period tended to focus on the Nestlé boycott and the hearings held over the marketing of breastmilk substitutes. For example, in addition to the December 6, 1981, article previously mentioned, a May 24, 1978, article in the *Washington Post* on the "Role of Infant-Formula Makers in Developing Nations" includes extensive history and coverage of the formula debates, a description of the Kennedy senate hearings on the issue, a discussion of the Interfaith Committee for Corporate Responsibility, and commentary on the ensuing problems in developing nations when we fail to put industry checks in place (Brown; Solomon "Controversy").

A series of four articles from May 19 through December 6, 1981, in the *New York Times* titled "The Diplomacy of Mother's Milk" address the fact that the Reagan administration stood alone against 150 countries when the World Health Assembly voted on the marketing code (Brozan; "Courageous"; "Diplomacy"; Solomon "Controversy"). This is a rare instance, reminding us that governmental policy plays a large role here. Adding to the discussion, a June 19 *Washington Post* article of the same year titled "Improper Use of Baby Formula Not Only an Overseas Problem" notes that not only is formula a problem in developing nations, but that "as many as 5000 deaths per year could be prevented with breastfeeding in the United States," and that the abandoning of infant formula could save 380 million in healthcare costs (Bonner). This article, significantly, also is the first to bring race into the equation, suggesting that many black women, who have to work outside the home more often than white women, more frequently find that they do not have a choice about how to feed their infants.

BREASTFEEDING AND ENVIRONMENTAL ADVOCATES: FRIENDS OR FOES?

While I fully support and celebrate the vast majority of this extensive and valuable work, I have discovered in my research another part of these efforts that has more disturbing ramifications. Because rates have been so low, part of the breastfeeding advocacy community has, at times, been so single-minded in its efforts to promote breastfeeding that they have attempted to suppress discussion of breastmilk toxicity, in effect censoring and silencing those who might raise such issues. If we think back to the medicalization of birthing and infant care, the takeover of these activities by the burgeoning medical profession in the later decades of the nineteenth century and the early decades of the twentieth century, the reign of "scientific motherhood," and the playing out of the women's movement, we see how certain supremacist gestures determined that a whole trail of events would lead, first to the drastic falling off of the numbers of women nursing, and then to the silencing of discussions of breastmilk contamination. I believe that advocacy groups can do the most good by acknowledging the reality of the environmental contamination of breastmilk—but in a way that does not prompt women to turn to formula, thinking this would constitute the healthiest approach to the dilemma of living in an atmosphere subjected to environmental pollutants.

To trace the history of attention to the environmental contamination of breastmilk from the perspective of the advocacy groups, we find an early silence, along with a reluctance to acknowledge the reality of breastmilk contamination. Indeed, a quick review of early advocacy literature reveals a marked absence of the topic. Even today, when I surveyed all the books available in my local library shelves on nursing, I found that the environmental contamination of breastmilk is rarely mentioned, and that when authors do broach the subject, they tend only to mention it in passing. Harrison notes that while we might expect "outrage or denial" from these groups, the "overwhelming reaction is denial," with some voices suggesting that allegations of contamination are "the work of formula companies seeking to discourage breastfeeding" (49, footnote 40).

Another common early reaction was for advocacy groups, on acknowledging breastmilk toxicity, to point out that formula and cow's milk also contain contaminants—a reasonable response, so long as it is lodged in the spirit of providing information rather than as a defensive retort. For example, Baby Milk Action, the British arm of the International Breast-feeding Advocacy group (IBFAN), repeatedly ran an article titled, first, "Scare Stories," then "Scare Stories Again," taking issue with those who raise the possibility of breastmilk contamination and accusing them of using fear tactics to frighten women away from nursing. The group stated, in 1997, after noting that "scares about contaminants in breastmilk may have a damaging effect on health workers and public support for breastfeeding," that "the baby food industry has capitalized on mother's insecurities, exaggerating the qualities of artificial milk and minimizing their known and proven risks" ("Good News"). It continues, using its own fear tactics, pointing out the dangers of using formula; "Artificial milks have been shown to contain high levels of aluminum, lead, and other heavy metals and in the last year alone concern has been raised about salmonella, phytoestrogen and phthalate contamination" ("Good News"). At other points, the group seems intent on denying, or at least deflecting attention from breastmilk contamination, reiterating that "contaminants also impact on artificial feeding," noting that formula itself often poses problems, and that some plastic bottles used for feeding infants formula have been shown to be risky because of the threat of exposure to Bisphenol A (BPA), "an industrial chemical used to manufacture polycarbonate and other plastic items" ("Contaminants"). If we follow the trajectory of attention by this group, we observe a change over time. For example, one later report notes that "the level of exposure in a bottle-fed infant is less than the tolerable daily intake, but greater than the quantity found to cause effects in studies on animals" ("Contaminants"). By this point, in 2000, IBFAN had been talking about contaminants from various sources and exposures to breast and bottle-fed babies for quite some time; the tone, which by then had lost all of its earlier defensiveness, had become savvy and much more confident of readers' ability to read between the lines. For example, exposures that are "less than the tolerable daily intake"—set by policymakers often responding to pressures by industry—but that are "greater than the quantity found to cause effects in studies in animals" are by no means meant to be taken as simply "safe."

The chronology—for this group and others—involves a period of silence around toxicity, and then resistance and defensiveness, followed by greater awareness, acknowledgment of toxicity in all forms of infant feeding, and, finally, a resourcefulness about the issue of environmental contaminants, as illustrated by the fact that Baby Milk Action now devotes an entire section of its website to "Past Press Releases and Articles on Contaminants and Infant Feeding." By 1997, Baby Milk Action issued a press release in which it acknowledged the presence of toxins in human breastmilk ("Good News"). Since the purpose of the release was to note new findings that PCBs and dioxin levels had fallen—at least in Europe—it could acknowledge the presence of carcinogens, yet give the news a positive spin, stating that "although the PCB levels in human milk still exist," "the news illustrates the effectiveness of campaigning on environmental issues"—a surprising and welcome development given the early tendency for many advocacy groups to treat environmental groups as enemies. Another development in this trend of advocacy groups coming together with environmental groups also took place in 1997.

Here, we see an advocacy group articulating the same message that some environmental groups had brought forth—that breastfeeding presents much less of a burden on earth's carrying capacity than bottle-feeding—thus approaching the problem of low breastfeeding rates from the perspective of the relative environmental implications of the two feeding methods. For example, Infact Canada, in "Breastmilk: The Perfect Renewable Resource" announces that "it is appropriate for us to consider the ecological impact of infant feeding practices." Indeed, the group combines toxics issues with sustainability issues, noting that "with continued depletion of rain forests, damage to the ozone layer, and chemical toxicants in our air, food, water and soil, breastfeeding with all its benefits, ranging from health, to social, to economical, is without question the only safe, sustainable and environmentally suitable means to feed babies." In addition, Andrew Radford, of Baby Milk Action, observes in "The Ecological Impact of Bottle Feeding" that breastmilk "is one of the few foodstuffs which is produced and delivered to the consumer without any pollution, unnecessary packaging or waste" (1). Articulating what is perhaps one of the clearest examples of breastfeeding advocates aligning themselves with environmental causes, he offers a deeply visual image of the folly of failing, as Carson warned, to see the big picture: "giving up breastfeeding in response to dioxin levels is self-defeating as artificial baby milks contain high levels of [pollutants] . . . moreover," he continues, "a decision in favor of artificial milk will lead to increased pollution and dioxin levels." (2). At another point, Radford, rather than taking a stab at environmentalists for scaring women and thus running the risk of thereby promoting formula-use, as some breastfeeding advocates have done through the years, points to how formula-producing industries constitute a problem for both groups: "The undermining of breastfeeding is the destruction of a natural resource and should therefore be seen in the same light as logging in the rainforests or overfishing our seas and rivers." Clearly, breastmilk advocates and environmental advocates have many reasons to serve as allies. But what Radford and other advocates repeatedly fail to do is to address the fact that this is not an all-or-nothing situation. While one could not deny that formula production and use burden and pollute the earth more than nursing does, it is important not to simply demonize all formula production. Why? Because artificial infant food can save lives as well as ease the burden on mothers at times when nursing proves impossible or problematic.

If 1997 marked a year in which breastmilk advocates and environmental advocates seemed to come together, 1999 was one in which early contentiousness again flared up. Some factions of the breastfeeding advocacy community have taken environmentalists to task for raising anxiety around breastmilk contamination and for purportedly using the issue to promote awareness of environmental degradation—even at the cost of scaring women away from breastfeeding. Baby Milk Action, in 1999, took the World Wide Fund for Nature (WWF) to task for its coverage of breastmilk contamination. Stating that "a report by the World Wide Fund for Nature (WWF) was published at the launch of World Breastfeeding Week and created scare stories around the world," the article, called "Scare Stories Again," complains that while WWF "wanted to highlight the risk of environmental contaminants . . . breastfeeding scare stories resulted in many countries." Critiquing WWF for how they framed the issue, the advocacy group's most recent approach has been to address warnings of contamination, but to highlight certain handlings of the issue. Noting that WWF had published a review of research, they argued that "perhaps WWF can learn from UNICEF who conducted a similar review of existing research in 1997," coverage that "did not look at breastfeeding in isolation," as WWF purportedly did. IBFAN argues that WWF cited a study estimating that "about three days of life expectancy would be lost because of cancer attributable to contaminant exposure through breastmilk," but did not mention studies showing that "the decreased life expectancy from not breastfeeding was about 70 days."

A comparison of WWF's "Chemical Trespass: A Toxic Legacy" and UNICEF's "Breastfeeding and Environmental Contamination" illustrates that both privilege terms such as "despite . . . ," "still . . .," and the weighing of "risks"—characteristic of later coverage of environmental toxicity—to emphasize that while environmental toxicity threatens the healthfulness of human breastmilk, research *still* confirms the comparative benefits of breastmilk over formula (Lyons). Both strongly endorse breastfeeding, even in the more extreme cases of contamination. WWF's section titled "Breastfeeding should still be encouraged" argues that "there is convincing evidence of the benefits of breastmilk to the overall health and development of the infant," and their contention that breastmilk "is the ideal nutrient" and "should certainly still be encouraged" compares with UNICEF's. While Baby Milk Action complains that WWF "look[s] at breastfeeding in isolation," a comparison of the two framing devices suggests that what Baby Milk Action fears is that WWF's positioning of the issue, following Rachel Carson and her successors, as "a postwar chemical revolution" that has affected wildlife and now impacts "the human race," and as an issue that demands immediate attention, may, "despite" its staunch endorsement of breastfeeding, bring women to such a degree of anxiety that they choose formula (3).

UNICEF's paper, in contrast, offers commentary that could be construed as undermining claims that the situation is a cause for alarm or for action. Its opening page contains, in a boldfaced box, the comments: "The presence of chemicals in breastmilk must be seen in perspective. Because a chemical has been detected and measured, it does not mean that it is necessarily harmful to the infant in the quantities consumed" (World Health Organization, "Principles for Evaluating"). It continues: "Despite the concentrations of pesticides found in human milk, no major studies have demonstrated that the presence of pesticide concentrations have led to adverse health outcomes in the children exposed through breastfeeding" (National Research Council, "Pesticides"). While the focus only on "pesticides"—a somewhat loose term—is misleading, the blanketing commentary on studies showing "no adverse health outcomes" could also be seen as misleading. All this is saying is that studies on effects on humans had not, at that time, been done—not that studies have shown no adverse effects. And one could question such a contention. WWF, in contrast, notes "potential effects," such as human studies demonstrating links with "reduced intelligence and/or behavioural effects" and animal experiments pointing to "endocrine disruption, abnormal reproductive system development, abnormal nervous system development, and other developmental abnormalities" (4–5). Suggesting extensive effects on shrinking global biodiversity and implications of exposures to sensitive populations such as the Inuit, who live far from contamination sources, and Mexican agricultural workers exposed to pesticides, WWF advances environmental justice causes, while articulating "the need for global agreements to reduce or eliminate discharges . . . of the most persistent and toxic substances which can cross international boundaries"(7).

Following Carson, it thus challenges us to see that even the seemingly smallest and most innocuous actions can have grave and far-reaching consequences. Commenting that "WWF has played a key role in raising the alarm about problems associated with EDCs," IBFAN's discussion brings to light the fact that WWF could have been more careful about their participation in situations in which "Baby Milk Action received anxious calls from health workers working to protect breastfeeding and infant health, in countries such as Saudi Arabia, Malaysia and India" (1). But in reprimanding WWF for their approach, suggesting that "perhaps WWF can learn from UNICEF" on this issue, Baby Milk Action undermines the kind of back-and-forth discussion that we need, and fails to note that UNICEF's approach could also be strengthened by attention to some of WWF's discussion.

While a wide array of breastfeeding advocates and advocacy groups offer responsible, articulate, and persuasive arguments in support of breastfeeding, some of these groups adopt what could be viewed as more outright anti-environmentalist rhetoric. They fear that bringing any attention to breastmilk contamination will encourage more formula use, and they do not trust caregivers to choose nursing if made aware of contamination. For example, in *Milk, Money, and Madness*, Baumslag and Michels, two prominent breastfeeding advocates who have written passionately and knowledgeably about nursing, comment that "the fear that breastmilk is a reservoir of concentrated poisons is not new and is one more effort to discredit breastmilk" (96). They go on to state that "there has always been speculation that artificial

feeding was a way of saving children from their mother's contaminated milk."
While consideration of the history surrounding formula production and use
suggests that such a claim is justified, the direction in which they choose to
take this suggests a certain irresponsibility, defensiveness, and fear. Mirroring
anti-environmentalists who seek to discredit claims of environmental problems, they set up their argument to articulate a view that characterizes some
of the more extreme advocacy rhetoric. They continue: "Claims of contamination must be looked at in light of their sources: environmentalists often
exaggerate the degree of contamination to boost their case for greater safe-
guards, and formula companies freely spread fears of toxins in the hopes that
women will choose formula as a way of erring on the side of safety" (96). In
positioning environmentalist claims of contamination as a ploy calculated to
play on fears of unsuspecting citizens, and as a tactic aligned with the "bad
guys" in this drama—formula industries—they set their cause in opposition to
that of potential allies. Such rhetoric is particularly upsetting, given that *Milk,
Money, and Madness* offers an engaging read with articulate and useful commentary. While many environmentalist groups have exhibited a wariness of
bringing attention to breastmilk toxicity for many years, perhaps in response
to such characterizations, a few groups have begun to address the issue.

Indeed, the latest phase of interaction between environmental groups
and advocates seems to be marked by a renewed openness about acknowledging the environmental contamination of breastmilk, plus a new focus—one
that looks more pointedly at the issue from an environmental justice angle.
The Natural Resources Defense Council (NRDC), for example, recently—
in 2001—took up a "Healthy Milk, Healthy Baby: Chemical Pollution in
Mother's Milk" campaign. Their wording attests to the degree to which the
very notion of breastmilk contamination remains a surprise to most people.
Noting that new mothers will describe a wide range of feelings about nursing,
they comment "What very few of them have reason to suspect is that in
addition to all the positives about breastfeeding, scientists are now becoming
aware of a small but troubling negative: small doses of chemical pollution are
invading mother's milk" ("Chemicals"). Anticipating confusion from mothers and backlash from breastfeeding advocacy groups, they announce: "It's
not time to panic and not time to stop nursing." They handle the question
of recommendations by inserting the word "almost": "In almost all cases, the
health benefits of nursing far outweigh the potential problems from POPs."
Here, persistent organochlorine pollutants (POPs) emerge as the threat, replacing DDT and PCBs associated more fully with Carson's day and beyond,
reminding us that as we phase out certain chemicals, others emerge. The
NRDC takes the opportunity to suggest other recommendations, placing the
problem within a larger context: "While it may not be time to panic," they
claim, "It is time to learn more about the problem, and ultimately, to take
social and political action to protect our children's children from facing a
more difficult choice when it's their time to care for a newborn."

Other environmental groups who did not address the issue of environmentally contaminated breastmilk early on have taken it up later, as part of their approach to larger environmental dilemmas. For example, as previously noted, the Sierra Club, who had, for many years, considered toxics issues more generally, but did not focus on breastmilk contamination specifically, mentions POPs and indigenous health in the 2000 *EJ Times*. Noting that "that living elixir of human love and bonding, mother's milk, now imparts poison along with protein to the newborn," the article explains that POPs "resist chemical, microbial, and physical decomposition" and thus persist in the environment (Sinclair 3). Offering readers significant detail, it points out that these chemicals remain and "reaccumulate" by two mechanisms— "geoaccumulating," or "repeatedly volatizing and then being swept up into the prevailing global air currents, which transport them to the poles," or bioaccumulating "through the food chain to levels that are toxic to humans and to animals near the top" (Sinclair 3). Such detail and frankness in confronting the situation bring us, in many ways, back to Carson and an approach that, offering richly and carefully worded explanations of scientific concepts, trusts citizens to make their own decisions.

THE ROLE OF THE PRESS IN DISSEMINATING STORIES OF BREASTMILK TOXICITY

In several instances, the mainstream press has done a commendable job of disseminating information about breastmilk toxicity. Indeed, several articles through the years have been articulate and forthright in following in Rachel Carson's footsteps, explaining complex processes such as bioaccumulation, alerting readers to how babies occupy a precarious position at the top of the food chain, or informing us about developments in policy that affect mother's milk. For example, an article published in February 1980 in the *Washington Post*, titled "Firms Exporting Products Banned as Risks in US" looks at how each year the United States "exports millions of pounds of chemicals banned from domestic use because they are too hazardous" and then turns around and imports such chemicals—including aldrin, dieldrin, heptachlor, and chlordane (all discussed in *Silent Spring*)—on "cacao from Ecuador, coffee from Costa Rica, sugar and tea from India," and many other sources of contamination (Hornblower). The article quotes the General Accounting Office's conclusion that "a large portion of food imported into the United States may in fact contain unsafe pesticide residues." And it explains some ramifications that ensue for those we often forget, who actually work with the substances: "In some foreign countries pesticides known or suspected of causing cancer, birth defects and gene mutations are carelessly or excessively used." It notes the shameful occurrence that "even pesticides approved for use here can be dangerous when exported and used by illiterate farmhands." This problem is often exacerbated by the fact that such instructions are printed only in English.

In the vein of Carson, the article treats readers to anecdotes rich in detail, putting a human face on the issue. It explains that "in 1976, at least five Pakistanis died and 2,900 became ill from the common pesticide malathion" (Hornblower). The workers, who had not received instructions on proper handling, had mixed the chemical with their bare hands, washed the spraying equipment in local water supplies, and "spilled the pesticide— which can be absorbed through the skin—in areas where barefoot children played." Significantly, the writer of the piece, Margot Hornblower, choose to begin with a story of how "in the poverty-stricken countryside of Guatemala, mother's milk is contaminated with DDT," and then discusses how the "export of hazardous products and industries is becoming a major controversy here and abroad," with some poor countries acting as "willing consumers" and others complaining of "being used as a dumping ground by U.S. companies." Noting that the sale of pesticides to developing countries, at the time, made up 40 percent of the $2.6-billion U.S. production of these chemicals, and that President Carter had, the preceding year, "set an important precedent with an executive order requiring environmental assessments for federal actions that could cause major pollution abroad," the article also reports that a proposal for an "executive order imposing more stringent controls over a wide variety of dangerous products" had reached a stage of "interagency conflict."

Interestingly, the *New York Times* offered similar reporting three years later, when it printed, from its "Editorial Desk," a letter titled "Poisons That Are Boomeranged Out of the US" (Regenstein). This piece also looks at how poisons that are banned here in the United States are shipped to other parts of the world and then reimported in the foods that we eat, thus endangering our health and rendering any policy banning such chemicals ineffectual. Again breastmilk plays a prominent role early in the article when readers are told that while DDT was banned in the US in 1972, it "was being found in over 99 percent of all human tissue samples taken in the U.S., as well as in breast milk and food, air, and water throughout the country." The piece reports that, sadly, "shortly after entering office in 1981, President Reagan revoked the modest export restrictions on such hazardous substances that had been implemented by executive order a month earlier." It also notes that that administration "has drafted plans to expedite the ability of U.S. companies to export banned products on the ground that lifting such restrictions placed U.S. exports at a competitive advantage.[8]

In May 1982, an article on the front page of the *Washington Post Magazine* cites the same author, Lewis Regenstein, in an article significantly titled "Did We Forget the Deadly Message of *Silent Spring?*" (Maxa). Noting that "twenty years ago, Rachel Carson warned the world of the dangers of man-made pollution," the author of the piece, Rudy Maxa, quotes Regenstein, who "fears Carson's overall message was lost."[9] Regenstein points out that, in Carson's day, "We were just detecting what appeared to be a rise in the cancer rate. Now," he exclaims, "it's an epidemic." Indeed, "One out of every four Americans living today will get cancer." He continues: "Last year 420,000 people died from it, more than the number of Americans who died in the battles of WWII, the Korean War and Vietnam combined." Cigarette smoking accounted for only about a third of those deaths, while the response that people are just living longer, and thus dying of cancer at higher rates, does not hold up, given that cancer is the leading cause of death for children aged one through ten. This brings the author to point to the current "high concentrations of cancer-causing chemicals in mothers' milk" and "imbalance of the food chain thanks to indiscriminate use of pesticides." The good news, proclaims the author, is that "people through their life styles" can minimize consumption of chemicals and can "demand that the government enforce laws on the books."

The article, which also mentions the publication of Regenstein's then new book on the subject of environmental pollution, *America the Poisoned*, is one of several through the years in which the press has played the beneficial role of alerting readers to new works on the subject. For example, in September 1980, the *New York Times* reported on the release of *At Highest Risk: Environmental Hazards to Young and Unborn Children* by Christopher Norwood; in April 2000, the Syracuse *Post-Standard* showcased Sandra Steingraber's book *Living Downstream* on the link between cancer and the environment, and her work on her then upcoming *Having Faith: An Ecologist's Journey into Motherhood*, which discusses at length what happens to the body during pregnancy, and the effects on the child, in utero and through breastmilk, from chemical pollutants (Klemsrud and Dooley; James). In many cases, the book review section provides more information than investigative reporting. Other noteworthy moments in the press's coverage of breastmilk contamination through the years include a 1987 *Washington Post* article that explains that a chemical banned for most uses, in large part because of findings that it had negatively affected human milk, has not been banned for termite use and so continues to pose a threat to human health (Weisskopf). We also find an article published in the *St. Petersburg Times* warning readers that the PCBs an infant drinks in her mother's milk will be passed on, not only to that infant's children, but also to her grandchildren, with accumulations intensifying with every step up the foodchain ("PCB Threat"). An article published in 1990 in the *Seattle Post*, titled "Study of Chemical Stirs Breast-Feeding Concerns," explains that while rates of exposures in some places are $16^{1}/_{2}$ times higher than the national standard, and that "although medical opinion is that the known benefits of breastfeeding outweigh any potential risks," "further contamination of human breast milk may lead to significant and unacceptable exposure," thus alerting readers that we cannot be complacent about such findings. These articles, in the aggregate, may suggest significant attention to the issue of breastmilk contamination, yet we must bear in mind that those few mentions of the issue took place over a span of years and were scattered throughout the United States. In addition, we must acknowledge the placement of such stories in individual newspapers—as too often the case with environmental news, such stories rarely appear on the front page or even in the A section. But that these stories have appeared at all is a good thing.

While the mainstream press has performed a valuable service in bringing attention to breastmilk contamination, at other times it has either not reported on breastmilk toxicity findings or has covered pertinent related issues without giving sufficient attention to breastmilk. For example, an article published in *The Record* about the release of an important breastmilk study—the Schector study of fifty mothers that showed a nursing infant will receive eighteen times the recommended lifetime dose of dioxins in the first year of life—was titled "No Alarm Over Dioxin in Human Milk." Although the opening of the piece, "Traces of toxic dioxins found in human milk should not cause alarm or stop women from breast-feeding their infants," conveys important information—that toxins are present, but that breastmilk still constitutes the healthiest food for infants—the title, proclaiming "No Alarm," is misleading, especially because many readers peruse newspaper headlines, without reading the full stories. Similarly, a 1989 *New York Times* piece on how Inuit women have the highest concentrations of chemical toxins in their breastmilk than any other people in the world, conveys the story without any mention of the process of bioaccumulation or any explanation of why it is that these people, far from polluting activities, have such high doses in their breastmilk ("High PCB Levels"). Such discussion would provide an opportunity for readers to confront the fact that pollution does not stay put, and that effects in one part of the world will, in time, become dispersed to other regions through the movement of wind, water, soil, animals, and food.[10]

While the press did cover, in May 1994, the leaked release of an EPA draft "Dioxin Reassessment"—a study noting that "populations receiving high-end exposures are approaching levels at which adverse effects have been reported in lab animals," another important study by Rogan, Blanton et al. called "Should the Presence of Carcinogens in Breastmilk Discourage Breastfeeding" received no press coverage (Rogan, Blanton et al.; Schneider).[11] And coverage of "Dioxin Reassessment," however, only hinted at "breastmilk as a source of exposure," while the absence of mention of the Rogan, Blanton study deprived the public of knowledge that breastfeeding shows reductions in mortality of 256 per 100,000 live births, greatly reduced mortality and morbidity ratios compared with formula-fed babies in developed and undeveloped nations (Harrison 43; Rogan, Blanton et al. 228–40).

THE PRESS ON TOXIC FISH

When we turn to the handling of fish contamination and discussions about various municipalities and governments deciding to issue advisories about eating certain fish species because of environmental toxicity, we find quite a bit of attention from the press. Indeed, beginning in 1976, when New York State began issuing periodic health advisories warning anglers, sports fishermen, and women about consumption of certain fish from certain waterways, the mainstream press has provided substantial coverage. What I find interesting is the position of contaminated fish in relation to contaminated breastmilk, and the way members of the press have chosen to address that relationship. While the mainstream press often appears reluctant to tackle the subject of toxic breastmilk on its own, as the main subject of investigation or reporting, it seems less reluctant to broach the subject of toxic fish—one of the main sources of exposure to breastfeeding women. This may be because journalists can choose, simply, to report on advisories, which usually mention nursing moms as a vulnerable group, without getting entangled in the more emotionally laden subject of toxic breastmilk. Articles will often note that advisories pertain more fully to pregnant women, nursing mothers, and children, yet readers rarely find discussions about the bioaccumulation of pollutants up the food chain and through generations. And rarely are there discussions or mention of why nursing moms constitute a sensitive population.

The environmental contamination of fish constitutes what many consider the greatest source of exposure that has an impact on the toxicity levels of breastmilk.[12] It offers an easily understood illustration of how bioaccumulation works: large fish such as swordfish and tuna, which eat smaller fish, thus occupying a rung higher on the food chain, are subjected to greater toxicity levels. Nevertheless, in many cases, when members of the press have chosen to focus on toxic fish, they have done so without even mentioning greatly elevated risk levels for infants ingesting the breastmilk of women who have, at some point in their lives, consumed these fish. For example, several *New York Times* articles either mention nursing moms as a sensitive population, but do not provide any detail, or do not even acknowledge nursing moms at all (Hanley "Jersey"; "Hazards of New York Fin Fish").

One telling occurrence in the history of attention to the environmental contamination of fish—and, by extension, breastmilk—is that while the mainstream press, as well as mainstream environmental groups, have devoted time and space to toxic fish, with mention in most articles of the fact that health advisories apply most particularly to "pregnant women, nursing mothers, women of child-bearing age and young children," advisories and discussions of toxic fish only recently began to address the fact that certain populations—Native Americans, immigrants, people of color, and people of limited income—fish, not just for sport, but as a subsistence activity, thus rendering these groups more vulnerable to environmental pollutants (Gargan; Lefferts). For example, while coverage of fish toxicity at least contains warnings directed at pregnant women, children, and nursing mothers in three *New York Times* articles from 1982—on June 2, October 24, and December 14, with similar coverage in other years—it was not until much later that "mainstream" media began to focus on heightened risks for these other sensitive populations

(except the Inuit). For example, a 1994 article in the *New York Times*—
"Efforts Revive River But Not Mohawk Life"—is one of the first to look at the
effects of pollutants on just one of many Native American groups affected by
environmental toxicity (Gargan; Hanley; "Hazards"; Schneider "Efforts").

DEVELOPMENTS IN RELATED CAMPAIGNS:
THE ENVIRONMENTAL JUSTICE MOVEMENT

The story of the growing awareness that breastmilk—and infant food in
general—are subject to environmental contamination in many ways leads up
to a particular climax that, surprisingly, takes us far away from Rachel Carson
and *Silent Spring*. We can locate that climax—or centerpiece—in what has
now come to be called the environmental justice movement. Recognizing
that toxic breastmilk impacts all of us—men and women, the elderly and
babies, heterosexuals, gays, lesbians, those transidentified, and others, and
people from all different racial, ethnic, and national groups—we must also
recognize that toxic breastmilk impacts some more than others. Many studies
have found that at this point in time the breastmilk of all women bears
traces of certain chemicals, such as DDT, PCBs, PBDEs, and that such traces
"have implications for breastfeeding and infant health" ("Breastfeeding and
Environmental"; Gladen, Jacobson, Jensen, Laug, Rogan, WHO "Levels,"
WHO "PCBs,"; WHO "Regional,"). The reality that all breastmilk shows
signs of environmental pollution means that even those who were never
breastfed and will never breastfeed bear marks of our collective decision-
making. Here, breastmilk serves as the canary in the mine; like animal sentinels, it tells us that if breastmilk is contaminated, so, too, are the rest of our
bodies—as well as our "environment" as a whole, including the plants, animals, water, and air that surround us.

But to trace the story of breastmilk pollution brings us to a truth that
has been formalized only in the last fifteen to twenty years—that some
groups of people live and work in "environments" that determine that their
bodies get subjected to greater amounts of environmental toxins than other
people's. And, consequently, some women's breastmilk has been shown in
multiple studies to be more heavily laden with environmental toxins. Hence,
their children and children's children have to contend with the effects of
such toxins on their bodies and in their lives. The realization of such inequitable distribution of the waste resulting from our short-sighted and harmful
environmental policies has come to be called the environmental justice
movement. A central moment for this movement came in 1987, when the
United Church of Christ Commission for Racial Justice published a report
called "Toxic Wastes and Race in the United States" that concluded: "Race
is a major factor in the presence of hazardous wastes in residential communities throughout the United States" (Hamilton 209; United Church). That
"three out of five African Americans and Mexican Americans live in com-

munities with uncontrolled toxic sites," that "75 percent of residents in rural areas of the Southwest are drinking pesticide-contaminated water," and that "more than 2 million tons of uranium trailings have been mined and dumped on Native American reservations" suggest that the term "environmental racism" names a very real problem (Hamilton 209). Like so many of the abuses noted by advocates in the women's movement, these abuses had been so insidious that, until they were named, they were given little notice.

If Carson brought pesticides, chemicals, and toxic substances to the table, arguing that these constitute threats, not only to birds, plants, fish, and ecosystems, but also to human beings who inhabit the land, recreate on and eat from the water, and breathe the air, environmental justice advocates helped us to see that some people's bodies are subjected to much higher rates of negative health outcomes. Before we could recognize that the environmental contamination of breastmilk affects some more than others, we had to get to a point where we understood that some populations live in areas and in ways that determine that they are more vulnerable to environmental toxicity. Until "environmental racism" was named, and identified, and studies demonstrating that certain populations find themselves more subjected to the negative effects of our environmental decision-making appeared, coverage of this aspect of breastfeeding did not surface. So, as we would expect, attention to the environmental justice part of the story did not emerge until fairly recently. While I do not want to go into detail about this aspect since chapter four looks at it in greater detail, I would like to sketch the basic contours of that story here.

What we find is a change in attention to the environmental contamination of breastmilk after the environmental justice movement entered the scene. Until this time, when most people thought of the "environmental" movement, they tended to conceptualize "green issues"—conservation, wildlife preservation, "wilderness conservation," global warming, keeping species from extinction, or preserving the rain forests. Often activists were adversarial to the perceived interests of working classes—farmers, fishermen, industry workers, and loggers. With the growth of the environmental justice movement, more people began to think of the term "environment" in an expanded light, that the "environment" is "where we live" and that urban environments and issues such as occupational safety, toxic waste disposal, trash incineration, the management of industrial and mining activities, and pesticide production and use could all and should be considered "environmental issues." Interestingly, as mentioned earlier, while the more established environmental groups tended, early on, to stay away from breastmilk contamination issues or to focus on related issues such as toxic fish, some environmental justice groups, which did not develop until later—in the late 1980s and early 1990s—did highlight breastmilk toxicity, perhaps because these groups were already attuned to issues of sexism, racism, ethnicity, and classism. For example, the Mother's Milk project, a collective of Mohawk

women who found that their breastmilk bore signs of environmental contamination at rates much higher than the general population, made it their
central cause to raise the public's awareness of the environmental pollution
of breastmilk. Perhaps because these people had long found themselves subjected to genocidal policies and practices and realized that their activism had
to face things head-on, they began campaigns against the health effects of
environmental pollutants earlier than many mainstream environmental groups.
Later, more environmental groups began to take note of the issue, and we
find more attention, and more detailed treatment of it in the mainstream
press. This may be due to the growing awareness of environmental issues in
general, as time passes. But it seems also to result from the way environmental justice advocacy has played out, and how this advocacy has engaged with
other movements, including feminist movements, environmental groups, and
breastfeeding advocacy groups. As the environmental justice movement began
to embrace the issue, seizing on the power and poignancy of the figure of
contaminated breastmilk, attention from other quarters began to take off,
perhaps because these groups felt pressured to weigh in since it had been
broached by EJ groups, perhaps because once attention had been raised,
these groups experienced the freedom to confront the issue without being
accused of being the instigators and thus the chief "fear-mongerers." An
example of the kind of attention one can now witness appears on Greenpeace's
website under "Health Effects: Toxics: Greenpeace USA." There, under
"POPs—Invisible Poisons," readers learn that "from the Arctic to the trop-
ics, POPs (persistent organic pollutants) can be found in the air, water and
food chain" and that "polar bears, whales and Inuit people living far away
from industry now carry extremely high levels of POPs in their body tissue
because POPs travel long distances by water, accumulating heavily in the
polar regions." Readers will also find a section on indigenous people and the
fact that "many scientists and tribal people consider POPs to be the greatest
threat to the long-term survival of Indigenous peoples."

THE SENSITIVE AND UNUSUAL NATURE OF BREASTMILK
ON A POLICY LEVEL AND AS A FLUID VITAL TO LIFE

Is there something about breastmilk itself or its environmental contamination that keeps us, as a collective, from recognizing the reality of it? I believe
so. First of all, if we look at the issue from a policy standpoint, we have to
recognize that breastmilk occupies a space different from formula or cow's
milk, which are regulated. Both cow's milk and formula must meet standards
set by the Food and Drug Administration (FDA), and must not contain
toxins over a certain level. But breastmilk does not get regulated by any
governing body; indeed, Rogan has pointed out that "breastmilk, if regulated
like infant formula, would commonly violate Food and Drug Administration
action levels for poisonous or deleterious substances in food and could not

be sold" (981). But the issue goes beyond the fact that breastmilk gets dis-seminated on a noncommercial framework, outside of a market economy, that we do not at present have policy frameworks that would regulate breastmilk contamination. The reality is that we do have policies in place, and could institute others that would work toward protecting breastmilk if we chose to focus on the issue.

Perhaps one reason for the lack of focus is that, as with all environ-mental health issues, it is difficult for people to conceptualize the effect that environmental pollution has on our bodies. We typically do not witness this "cause and effect" because, for the most part, environmental toxins affect our systems over time, with "visible" results only after the passage of years, if at all. Often, when people experience serious health consequences from envi-ronmental contamination (e.g., cancer, endocrine disruption, immunity sup-pression, or allergic reactions), these effects may not be seen as emanating from deleterious substances or the interaction of such substances in our environments and in our bodies. With more subtle conditions such as atten-tion deficit disorder, lowered cognitive functioning, or various behavioral disorders, which scientists are only beginning to understand as often result-ing from chemical reactions in our bodies, this is even more the case. Genetics, social setting, or "lifestyle choices" get blamed, and we fail to grasp the far-reaching effects of environmental pollution. But studies are begin-ning to suggest that environmental toxins do much more far-reaching dam-age than we have ever acknowledged (Gladen, Jacobson, Jensen, Rogan, Steingraber).

In addition, while scientists acknowledge that we have only begun to test the vast numbers of chemicals that we release into the environment, they also recognize that such chemicals have varied effects, depending on their combinations—and those, too, we have clearly just begun to measure. For the most part, we cannot see pollution in our water, our air, or in the foods we eat. Rivers and lakes rife with chemicals may look clear and "pristine"; green fields of hearty grains, bountiful fruits, and lush vegetables appear laden with healthy, wholesome nutrients—and they may be, except that they may also be laden with chemicals that have not been tested alone or in combinations. And soil soaked with harmful pollutants may appear dark, loamy, and rich. Fruits and vegetables sprayed with pesticides often have also been treated with chemicals that preserve them, help them transport better, or give them a brighter color or a more uniform appearance, thus making them "look better" or bigger than organic foods. And we have grown accustomed to this look. In addition, many foods that tend to be more subject to environmental pollutants seem, from our cultural perspective, to be "healthy," "hearty," or "homey" foods, including tuna fish, which we associate with the innocence of our third-grade lunches, or whole milk, ice cream, and other dairy products, seen as wholesome and good, swordfish, reflecting elegant dining, and meats and cheeses, the staple of good old American life.

With some environmental effects, one has a focusing event to latch onto: frogs born with three legs, the catastrophes at Three Mile Island and Chernobyl, and other similar events. We can feel the effects of urban sprawl when we drive about in our cars, but when it comes to the environmental contamination of breastmilk there is no focusing event. As Tom Birkland points out, these focusing events act powerfully to usher new issues into the collective consciousness and put them on the political agenda. He notes, "Sudden, dramatic and often harmful, focusing events give pro-change groups significant advantages" in bringing issues to the forefront of the political agenda (56, quoted in Harrison 46). Without such events people can overlook effects—even if studies have demonstrated that these effects are significant. Harrison points out: "It is beyond question that if there had been an epidemic of babies becoming noticeably sick from drinking contaminated breastmilk, this issue would have assumed dramatic proportions" (46). But clearly such events are not necessary for an issue to hit the radar screen. We can compare the reaction to the use of alar, when mothers became aware that the pesticide for spraying apples was highly toxic and was subjecting toddlers' favorite beverage—apple juice—to contamination. The alar case had no focusing event other than the release of studies about its toxicity, yet mothers lobbied hard and moved the apple producers to stop using the pesticide (Commoner 62–63). This incident corroborates that a focusing event is not absolutely necessary for bringing attention to toxicity issues. It gives us hope.

So why is breastmilk different? Is it because fewer mothers feed their children breastmilk than apple juice? Perhaps, but important to our discussion of the environmental contamination of breastmilk is the acknowledgment that breastmilk occupies a sensitive and unusual place in our thinking. More than any other substance or activity—even apple juice, a popular drink for toddlers and young children—it represents tenderness, innocence, sweetness, and love. It is the epitome of human connection, reminding us that ideally—even though things do not always play out this way—the love of a mother toward her child stands as the quintessential model of nurturing selflessness. Symbolically, as an icon, it signals human connectedness, with the establishment of the bonds that lead to language use and allow us to connect with one another in the ways that make life meaningful. The rush of emotions I felt when I first nursed my son, and that returns every time I feed him (he is seven months old as I write this) reminds me that all the clichés about mother's love that, frankly, seemed sappy and exaggerated before I became a mother, now ring true and surprise me with their accuracy. And even for those who do not nurse, will not nurse, and never were nursed, the figure of breastmilk still speaks of the quintessential human connection.

It may be that because breastmilk occupies such a position in our thinking that it is too far of a leap to conceptualize the effects of toxic pollutants on our breastmilk. Perhaps it is simply too poignant to think about. Or perhaps, despite post-1960s sexual freedom, our prudishness keeps us from focusing on it. To conceptualize it is to realize that we, as a community, have made decisions collectively that mean that our most vulnerable— our babies, and ultimately all future generations—are being subjected to harmful effects that impact their physical, cognitive, and intellectual development. Harrison's research into dioxin-contaminated breastmilk suggests that this is so. In part, she found an unwillingness on the part of some environmentalists to raise the issue because of the fear that doing so "might cause undue alarm and prompt some women to discontinue breastfeeding" (53). She quotes an environmentalist who explained "There was an incredible fear within Greenpeace. People always said 'you can't *not* breastfeed' and 'when we look at the data, it's really frightening. We are all very concerned that what we do doesn't turn into a massive marketing campaign for formula' " (53). On the other hand, feelings about the iconographic position of breastmilk play a big role in keeping the issue hidden. Harrison described how the male staff member of Greenpeace who had developed the campaign around the issue noted what happened when he tried to launch that campaign. He recalled "I really hit a raw nerve. There was a big reaction [among Greenpeace staff, especially the women saying] 'We don't want people to panic.'" The organizer said, "I hit the roof. That's paternalistic and condescending. If people's kids are at risk, they *should* get worried" (53). Harrison notes that "their comments indicate that their decision not to focus on breast milk were more than just strategic anticipation of backlash or self-restraint lest there be unintended consequences, but rather highly personal and emotional reactions" (53). Clearly, for these environmentalists, and probably for many others, the heavily laden emotional pull of breastmilk enters the picture and has contributed to keeping breastmilk contamination under wraps.

The wave of new studies of breastmilk contamination—of polybrominated diphenyl ethers (PBDEs)—has brought about a marked shift in attention to the issue. These studies, which indicate that American women's breastmilk contains surprisingly high levels of PBDEs, have garnered widespread, if fleeting coverage in many major newpapers and other media sources. While I have suggested that the focus on to the environmental contamination of breastmilk has assumed a circuitous path through the years, sometimes bringing a fair amount of attention, sometimes hardly getting noticed, this latest flurry is surprising in just how widespread it has become. Many sources, for example, *Pensacola News Journal's* "Breastmilk Best, Despite Toxins," offer headlines similar to those found at earlier junctures. But others, for example, Scripps Howard News Service's "An Emerging Health Crisis Involving Flame Retardants," indicate the degree to which these latest findings present a new chapter. Indeed, coverage almost unanimously reflects past trends and lessons, with journalists, for the most part, careful to highlight the old adage that "breast is still best."

But the wide coverage of the issue—it appeared in at least twenty-four sources from late September 2003 through late January 2004—and some specifics associated with the findings suggest where the difference lies. Despite claims that breastmilk remains best, a certain fear seems to be associated with this new round of attention. This fear appears to stem from several factors. First, PBDEs contaminate breastmilk at much higher rates here than in other countries, suggesting the need the United States wake up to the realization that our reluctance to ban chemicals phased out in Europe and elsewhere has real consequences. Second, scientists remain somewhat baffled as to the routes of exposures to people; while we know manufacturers use these flame retardants in a wide range of products, including computers, furniture foams, textiles, and consumer electonics, and recognize that these chemicals leach out continuously, the question of the exact route of exposure for PBDEs is still under investigation (Sterens). A third reason for this different handling is that more people—or at least more journalists and media personnel—seem to be beginning to realize that governmental regulation under the Toxic Substance Control Act is so weak that "there is no legal requirement to test most chemicals for health effects at any stage of production, marketing, or use," and that this makes us vulnerable to contamination from a staggering number of sources. This situation becomes more alarming when we confront the lack of testing, not only of individual chemicals, but also of these chemicals in combination with other chemical compounds. While we can point to specific regulations and the fact that we need to implement greater controls, the larger problem remains—that underlying systemic issues, the very structures of our rendition of capitalism in the United States, work to protect corporate interests. This incurs several costs, one of which is to human health. Quite simply, until we privilege the precautionary principle, which advocates banning chemicals until they have been tested alone and in combination—until we embrace health over protection—we will not have done enough. A hopeful note is that some recent legislation has been passed in the effort to phase out some PBDEs.

To reach a point at which the environmental contamination of breastmilk is no longer a menace is to get to the point where the environment is free of toxins, we no longer carry out polluting activities, and we have cleaned up the toxic remains of past polluting activities. To some, arrival at this point seems so utopian as to be unimaginable; to others, we seem so distant from this point that to confront the problem of breastmilk contamination is to confront just how far we have to go. This could be said of all environmental health issues, however. If breastmilk is contaminated, so, too, are cow's milk and various other ingredients in infant formulas; we cannot let the typical juxtaposition of "breast versus bottle" that has become so looming in the context of questions about formula advertising or individual mother's choices in balancing work and motherhood hide from us the reality that when environmental pollutants affect our future generations, the time has come for us to call for changes.

THREE

BREAST FETISHIZATION, BREAST CANCER, AND BREAST AUGMENTATION

THE CURIOUS OMISSIONS OF BREASTFEEDING AND BREASTMILK CONTAMINATION AS SIGNIFICANT FEMINIST ISSUES

A LACK OF FEMINIST ATTENTION

A cursory survey suggests that, by and large, self-identified feminists, those who position their feminism as central to their self-conceptions, do not view breastfeeding or the environmental contamination of breastmilk as significant feminist issues. While some recent attention indicates that this attitude may be shifting somewhat, the reality is that feminists have not focused on child sustenance issues or positioned them within a larger feminist framework. Some readers might find such a claim awkward and problematic—after all, many La Leche Leaguers identify as feminists and clearly embrace breastfeeding. And the leftist-leaning, organic advocating pop magazine *Mothering* has from time to time covered breastfeeding, with a recent issue focusing on the environmental contamination of breastmilk.[1] I suspect that many of its readers consider themselves feminists. In addition, environmental justice activists' literature has focused on breastmilk contamination. But although these exhibit feminist praxis, they do not self-identify as "feminist," and so I treat them as occupying a position somewhat different from that of more academically oriented feminists.

One problem involves the term "feminism" and the difficulties surrounding our attempts to define it. In some respects, more academically oriented feminists have claimed the term, seeing sexism, racism, ageism, colonialism, heterosexim, and speciesism, as well as other oppressive approaches as part of a vast network of privilege that serves the interests of Power. Feminist critic Suzanne Pharr's identification of a defined "norm"—white, Christian, heterosexual, able-bodied, monied men who wield power and privilege—works well in beginning to address the "patriarchal" system located as the problem but has the added complication that most feminists would agree that men who meet the requirements of the "norm" can themselves be feminists (53). The picture is further muddled because many who uphold what we might call "feminist" ideals reject the term. In the latter portions of this chapter, I will consider environmental justice activists, many of whom, according to critic Laura Pulido, "often choose *not* to emphasize their female identities, but rather define themselves in terms of race and place" (Pulido 19). Others tend to support what I might label "feminist" perspectives, but may not embrace the term, or may, in some cases, clearly reject it. For example, my mother, sisters, and some of my students uphold many of the ideals I associate with feminisms; they support equal pay for equal work; they advocate for many social justice issues; they are pro-choice; and they believe that having more women in decision-making positions would be a good thing. But, because they associate feminists with a certain anti-man and anti-traditional positionality that seems too angry, too anti-establishment, and too leftist, they do not embrace the term. And several theorists, in particular those of color, have pointed to problems that many women of color have with embracing the label, most prominent of which is the view that feminists, for many years, tended to approach things from a white, middle-upper class, heterosexist, and colonizing perspective (Davis; Lorde "Age"). Given the breadth of feminist viewpoints, it makes sense to talk about "feminisms" in the plural, and to use the term loosely, allowing it to encompass a wide range of anti-privilege visions.

So while I will start out focusing on academic feminisms, I do recognize that many whose interests may lie outside of academic approaches also either see themselves as feminists or reject the term but advocate for positions widely held to be feminist. I will also address (in)attention to breastmilk by these women. Contemplating the lack of feminist consideration of infant feeding and breastmilk toxicity, I find that this is a curious and surprising phenomenon given that feminists have prioritized several clearly related issues, including reproductive concerns, the sexual division of labor, and breasts and their embodiment in a patriarchal, heterosexist culture—or issues that I believe connect in specific ways to infant feeding. Feminists have critiqued cultural models that teach women to distrust our physical embodiment and see our bodies in narrow ways, as sexual objects in the service of male heterosexist fantasies, but have not tended to venture into the fields of breastfeeding and child nourishment. Why the focus on the sexualization of breasts, or on breast cancer, but not on breastfeeding?

A quick review of (in)attention to infant feeding in feminist journals corroborates this picture. In response to a quest for articles about "breastmilk" or "breastfeeding," "mother" or "motherhood," and "feminism(s)" in a selection of seven scholarly women's studies journals identified by the National Women's Studies Task Force, I found only a limited number.[2] While more have to do with the concept, politics, and/ or social-cultural construction of motherhood than with infant feeding, those topics also yielded limited results. For example, motherhood and infant feeding offered the following: *NWSA*—10 on "motherhood," none specifically on breastfeeding; *Women's Studies Quarterly*—11 on "motherhood," none on breastfeeding; *Feminist Studies*—37 on "motherhood," 2 on breastfeeding; *Signs*—64 on "motherhood," 1 explicitly on breastfeeding; *Gender and Society*—41 on "motherhood, 5 on infant feeding; *Women's Studies International Forum*—33 on "motherhood," none specifically on breastfeeding; *Women and Language*—17 on "motherhood," 3 on infant feeding. None of these journals included articles focusing on the environmental contamination of breastmilk or infant formula. MS. had 41 articles on "motherhood," with 3 explicitly on breastfeeding, while NOW's website (as an indicator of liberal feminisms' priorities) offered seven documents touching on breastfeeding, including a demand for "Greater Acceptance of and Access for Breastfeeding Mothers," a commentary on the American Academy of Pediatrics' recommendation that babies should be breastfed exclusively for 6 months and subsequently for at least one year, and resolutions and campaigns supporting pay equity, the creation of "women-friendly workplaces," and increased spending on childcare.[3]

Looking beyond those sources identified by the National Women's Studies Task Force, *Books in Print* offers 193 sources on infant feeding, including medical and science books (of which there are many), breastfeeding "how to" and advocacy books, (of which there are many), and social/political analyses (of which there are few); Amazon.com offers two selections (2002). Sandra Steingraber's *Having Faith: An Ecologist's Journey to Motherhood*, which does not focus on political, social justice, or feminist issues, and Kathryn Harrison's "Too Close to Home: Dioxin Contamination of Breastmilk and the Political Agenda," which concentrates on Canada and does not assume a feminist perspective, are rare in that they do focus specifically on the environmental contamination of breastmilk. Bernice Hausman's *Mother's Milk: Breastfeeding Controversies in America*, while self-identified as a feminist text, does not broach the environmental contamination of breastmilk. Christine Gross-Loh's commentary in *Mothering*, while not representative of more academically oriented feminisms, does focus on breastmilk contamination and begins to address some of the political aspects of this phenomenon, as does attention by a few others, including environmental justice activists, whom I will discuss more fully in chapter four. Although the limited scope of my search must be taken into account, the results do, however, suggest that the majority of self-identified feminists publishing in the United States do not regard infant feeding and breastmilk toxicity as matters worthy of extended commentary.

Not surprisingly, of the few feminist scholars who have written about infant feeding, several have noted how little feminist commentary exists. Penny Van Esterick, an anthropoligist by training, points out that "although breastfeeding is recognized as a women's issue, it is seldom framed as a feminist issue. In fact, it is most often ignored by feminist theorists" ("Breastfeeding" S42). Feminist scholar Linda Blum argues that "surprisingly little attention has been given to breastfeeding by feminist analysts, even in recent work focusing on the social construction of women's embodied experiences" (291). And social theorist and policy analyst Judith Galtry notes that "while pregnancy and childbirth have been the subject of intense feminist interest and debates, breastfeeding and its intersection with women's increasing participation in paid work has not been foregrounded either within feminist equality/ difference debates or within recent labor market analyses" ("Suckling" 1). Professor Bernice Hausman, whose book *Mother's Milk: Breastfeeding Controversies in American Culture* was recently published, echoes Galtry: "While the medical management of childbirth has received a lot of attention from feminist scholars, until very recently the medical management of infant feeding has received relatively little" (190). Anthropologist Vanessa Maher agrees, offering a partial answer: "In the 1960s and early 70s, maternity was not a prime feminist issue in the West. Women were more concerned with freeing themselves from childbearing and rearing than with realizing the potential of these roles as a female resource" (1). Noting that the 1980s brought "a widespread visibility to childbirth," she agrees with other critics that, even with increased attention to motherhood, infant feeding "did not share the limelight."

What strikes me most forcefully are the huge gap between the many books and articles approaching infant feeding as an isolated event, the attention lodged by discursive players who fail to position the issue within larger structural frameworks, and the relatively small amount of notice from those who might frame such issues within broader cultural paradigms and emerging cultural analyses. Although feminists devote much attention to violence against women and children, conceived either in literal terms or as systemic "violence," the violence inflicted on people without their consent or their knowledge—when their bodies become subjected to environmental toxicity through placental fluid or breastmilk—remains largely ignored.

After considering why feminists do not tend to embrace infant feeding and toxicity as significant feminist issues, and discussing some of the results of such an omission, I will survey some of the feminist attention to infant feeding that has been published and will argue that bringing the environmental contamination of breastmilk to the table, and allowing some of the work that has already been done to bear on explicitly feminist scholarship, models a way for feminisms to be more fully responsive.

Feminists have powerfully critiqued the ways in which patriarchal attitudes have attempted to situate the breast within popular culture. Many have worked to help us see the female breast not as an object to be fetishized by males, but as a vital part of living individuals, articulate, creative, thinking women who can shape events, forge policies and legislation, lead states, cities, countries, and families, and make life better for more people. Other feminists have looked at the fact that breast cancer rates have increased 40 percent overall since 1973 when the government first began tracking cases, and that childhood cancers have risen by one-third since 1950, and they call attention to the gender and age-based components at stake (Friend; Steingraber *Living* 38). They claim that something is going on here besides just coincidence, and point to environmental causes. They note that how we live and the choices we make have determined that with all types of cancer combined, the incidence of cancer in the United States rose 49.3 percent between 1950 and 1991, and continues to rise; with lung cancer excluded, overall rates have risen by 35 percent (Steingraber *Living* 40). Activists have also been responsible for the fact that breast cancer is currently one of the most studied diseases. This is just a tiny fraction of the kinds of work feminists have done to make this a healthier, more livable world. Imagine what power feminists could bring to infant feeding.

A feminist focus on this issue could help garner much needed attention to problems including the lack of institutional support for expressing breastmilk at work, childcare subsidies, and benefits that would ease the burden on childcare workers. It could foster awareness of the need for subsidies for breast pumps and lactation specialists, and could raise awareness around the lack of public support for breastfeeding in general. Feminists can help assess to the ecological impacts of formula production and use and dilemmas around HIV transmission through breastmilk. We can help address the problematic of differences in effects of corporatization, late-stage capitalism, and globalization on differently located women. Bringing feminist analyses to the table also highlights how questions about infant feeding and the environmental contamination of breastmilk involve systemic discrimination and oppression. It brings us to consider the ramifications on people's lives of infant feeding choices, for example on women of color who are more often employed to feed the children of "working" mothers and so denied, under our current system the opportunity to use breastmilk themselves—and who are disproportionately subjected to the environmental contamination of breastmilk here and abroad. As people who care about creating healthier and more just environments, we must work to overcome silences surrounding infant feeding, and to forge a vision that we can do the things that need to be done to reverse the unhealthy trends around breastmilk and infant sustenance that are currently in place.

FEMINIST SILENCE ON INFANT FEEDING: FROM PREOCCUPATION WITH THE SEXUAL DIVISION OF LABOR TO FIGHTS FOR REPRODUCTIVE FREEDOMS

Perhaps the primary reason feminists have not tended to address infant feeding as a significant feminist issue is that, historically, women's activism in the United States has been dedicated to dismantling the sexual division of labor, and many see infant feeding, particularly breastfeeding, as relegating women to the home or forcing them to bear the primary responsibility for childcare. Historically, feminists have been concerned with getting women out of the "private," "domestic" sphere and into the "public" domain, a move that allows women to fight for economic freedom, for political legitimacy, for a host of rights, and for having their voices heard. To influence decisions and vie for power and control over their lives, women have had to work hard to move away from being identified solely as wives and mothers. Having worked relentlessly to break the sexual division of labor and the residual effects of traditional social roles, many feminists may be wary of promoting breastfeeding and thus risking putting women "back" into domestic spaces, geographies defined by unremunerated work that often gets marked as "mundane" and "unimportant."

Linked to this is the fact that for many people the figure of the nursing woman invokes the "traditional" heterosexual nuclear family as well as work structures and values based on the privileged norms of our particular patriarchal, late-stage capitalistic culture (Blum; Carter; Galtry). Since "reproductive biology" has served in many cultural settings as the "explicit rationale" for patriarchal privilege, including social, economic, and political discrimination against women, feminists have sought to downplay markers of "reproductive difference," while also seeking to problematize "discursive parameters" that emphasize sameness as a "prerequisite to equality" (Galtry "Extending" 301; Gibson 1145–82). As feminist critic Pam Carter points out, breastfeeding "represents one of the central dilemmas of feminism: should women attempt to minimize gender differences as a path to liberation or should they embrace gender difference through fighting to remove the constraints placed on them by patriarchy and capitalism, thus becoming more 'truly' women?" (14). Having worked so hard to break the model whereby women have been shut out of certain professions and positions of power and remunerated on an unequal basis, many feminists appear cautious about embracing practices that run the risk of constituting a step backward. And the image of the nursing mother raises that danger, purportedly positioning women as solely responsible for the care of children.

In pointing to the traditional sexual division of labor coupled with the reluctance to risk putting women "back" into the world of nonremunerated and undervalued work as stumbling blocks preventing feminists from addressing infant feeding as a significant feminist issue, two issues need to be addressed. One is that historically and today, women of color and poor women of all ethnicities have been located differently than the privileged norm, and have more often had to work outside their own homes, so that the fight for release from the domestic sphere holds a different resonance for these women. Another issue that emerges, here, is that to see infant feeding to be associated with the traditional sexual division of labor is to recognize that, whether one chooses breastmilk or formula, work within the home or work outside the home, someone still must take on the role of feeding babies and toddlers, and most often, this duty falls to women—whether the biological mother or adoptive mother, or other caregiver.

Linked to the "sexual division of labor" issue is the fact that feminists have worked so hard to win reproductive freedoms—the right to "voluntary motherhood," as it was called in the early 1900s, to birth control, to safe, legal abortions—that these have taken priority over other reproductive issues, including concerns around pregnancy and lactation. As with the sexual division of labor, reproductive issues bring to the fore a set of supremacist gestures that need to be addressed. As Angela Davis so eloquently points out, while white women have historically embraced reproductive choice as a key feminist issue, women of color and poor women have had to fight a different set of issues, including sterilization abuse and coercive abortions (202–21). Fights for reproductive freedoms, which continue today, have come to be associated with feminist arguments, determining that those who identify as feminists but who are opposed to abortion sometimes feel silenced, and that other "reproductive" issues such as "pregnancy loss" and infant feeding are forgotten (Layne "Breaking," "In Search").

FEMINIST CONCERN OVER MATERNAL SACRIFICE AND CURTAILED AUTONOMY

Several feminists claim that they refrain from privileging breastfeeding as a feminist issue because they do not want to burden women with another duty or be proscriptive about how women should manage yet another aspect of their lives (Blum; Carter; Law). Many seem acutely aware that to add dictates about infant and child nourishment to the list of tasks mothers must perform to be "good" is to add to the cultural weight of controls on women, particularly their bodies. For some, it is to perpetuate cycles of blame and impose models that position women, and in particular certain women (i.e., those who do not breastfeed), as a "problem." Because poor women, young women, women of color, and less educated women tend to breastfeed at lower rates in the United States, we must be conscious of racist, classist, ageist, and other gestures that could emerge in the handling of breastfeeding as the more responsible choice. Many dictates tend to assume that "mothers" are fully responsible for childcare, without questioning the cultural embeddedness of the category "mother." For example, Pam Carter, focusing on Britain, argues that the "dominant construction of infant feeding" there involves "an irrational, if natural, woman who needs to be told again and again what to do" (1). Feminist critic Jules Law, articulating a similar concern, argues that "breastmilk and motherhood have become the symbolic vehicles for a shift in the burden of resources and responsibilities back on women" (441). She cites as evidence the bumper sticker "Affordable healthcare begins with breastfeeding," and claims that such dictates "shift the burden for nothing less than public health policy itself squarely on the shoulders of women" (441).

Recent arguments implying that public health problems such as child morbidity, disease, obesity, neurological disorders, immunological deficiency, or "world health" problems that get blamed on "overpopulation"—particularly in the third world—could be overcome if only all women would breastfeed put the onus on women to "save" the day, and characterize women as the "problem" when they do not breastfeed. While breastmilk can help address many of these concerns, to blame such problems on women serves to mask their real causes, which more often have to do with the inequitable policies and practices, skewed values, and supremacist assumptions that inform them. Such arguments, similar to backlash propaganda that blames the unraveling of "American culture" on feminists, clearly indicate a need to sort through the ways we frame these issues. On the other hand, Bernice Hausman's argument that in some cases "the implication is that women can or should resist such social presciptions [to breastfeed], which is part of a historical pattern of controlling women through ideological regulation of their repro-ductive activities," points to another danger—of choosing not to breastfeed merely for the sake of "resisting" such "ideological regulations" (204).

Discussions about the "burden" placed on women in privileging breastmilk over formula should, therefore, involve the recognition that because of various structural issues, breastfeeding can tend to be "burdensome" in the United States, at least more so than in many other countries. The popularity of the "breast" versus "bottle" construction, which may be linked, in part, to the positioning of "breast" versus "bottle" like the earlier "breast" versus "wet nurse," as basically equal choices that involve decisions about managing one's time and one's independence from one's child, also seems to play a role here.

Several feminists note that breastfeeding advocates' sole emphasis on the child may be problematic in that it contributes to martyrdom, self-sacrifice, and the dissolution of the mother's autonomy (Blum; Hausman 193; Law). We could say that La Leche Leaguers, with their more traditional notions of family, have "gotten there first" and set the terms of the debate. Indeed, when many think of an image of a nursing woman, they tend to associate her with a "Leaguer," and with more traditional social roles. In this respect it should be noted that while La Leche League formerly maintained the impor-tance of infants feeding at the breast, more recently the group has embraced the breast pump as a necessary and viable part of contemporary experience ("FAQ"). Although they have worked to expand the image of the breastfeeding mom, the more traditional rendition has, nevertheless, come to influence our conceptualization of lactating women. Hausman points to another problem: "the strategy of optimistic encouragement employed by La Leche . . . can make it seem as if a mother just hasn't tried hard enough when, for a variety of reasons outside her control, breastfeeding doesn't work out" (206). My own experience at one local La Leche League meeting made me realize that at least in some instances such strategies continue to operate. I met a woman who had had breast reduction surgery and struggled to con-tinue nursing on the advice of her pediatrician, even though her child had only gained back his birth weight during his first four weeks of life; her situation attests to the dangers of approaches that maintain the directive to use breastmilk, whatever the circumstances.

ESSENTIALISM/CONSTRUCTIVISM DEBATES

Another reason feminists have not fully embraced infant feeding involves the wariness, on the part of some, to foreground anything that essentializes woman, particularly at a time when many feminists are working to bring the culture at large into an awareness of how gender is culturally constructed. According to feminist scholar Linda Blum, while motherhood "re-creates the core paradox and core ambivalence" of equality versus difference dilemmas, in breastfeeding, "the socially constructed and the biological are inextricably intertwined" (291). This involves essentialism/constructivism debates in which the assumedly liberatory practice of calling attention to the social construction of gendered identities comes up against the practice of calling attention to the essentialized category of "women" so as to allow it to operate as the political grounding for feminist movements—also potentially liberatory. In other words, sometimes it benefits feminists to call attention to gender not as a "material," "God-given" category, but as a socially constructed one, as when feminists claim that women are no less suited to be scientists or doctors than men are.

But, on the other hand, for feminists to have a platform on which to base a movement, they must identify as a group that is essentially different from men. Breastfeeding, as a symbol of "essential" womanhood, throws this problematic into relief. Clearly, while we associate "essentialism/constructivism" debates with the 1980s and 1990s, they continue to plague feminist analyses. Accepting narrow, limiting biologism has supported narrow, limiting social roles for women. Despite several analyses that problematize the very notion of strictly binarized matrixes—for example, arguments such as Diana Fuss's *Essentially Speaking* or Guyatri Spivak's concept of "strategic essentialism," which deconstructs hegemonic binarization—vestiges of the problem still remain active in feminist discussions and linger in feminist approaches (Fraser; Fuss; Kristeva).

STRUCTURAL IMPEDIMENTS

If we look specifically at academic women, it is also important to examine the question of whether something within the academy either prevents these women from being able to manage breastfeeding or from recognizing it as a significant issue. And I suspect that the situations faced by these women parallel the situations faced by other professionals. Several articles discussed recently in the *Chronicle of Higher Education* suggest that policies surrounding paid leave for academic women, along with perceptions about how others might view them if they appeared to place family issues over careers, might very well play a role in the depriveging of breastfeeding by academic women (Evans C4; Fogg A10–13; Steele C4; Wilson 1–5). Indeed, current coverage in several studies and journals leads one to believe that the academy does not tend to offer a family-friendly atmosphere, particularly for academic women, and that academics tend to play down their family commitments.[4]

Noting that only 34 percent of private universities and 18 percent of public institutions offer paid leave beyond the six weeks typically authorized by doctors for postpartum recovery, one of the *Chronicle* articles states that academic women may not have sufficient time off to fully establish and maintain breastfeeding and then to manage the switch to expressing milk (Wilson). Based on a report called "Parental Leave in Academia," which draws on a national study of 168 institutions completed in 2001, the article notes that, taken together, only 26 percent of the universities studied had policies that offered some paid leave to parents beyond the six-week maternity leave. Pointing out that even when universities offer family-friendly policies, many women do not take advantage of them for fear of damaging their careers, the article implies that academic women who may be devoted to breastfeeding may choose to downplay it, in an attempt to distance themselves from perceptions that they may be privileging parenting over their academic careers. A separate study of 5,087 faculty members conducted by researchers at Pennsylvania State University's University Park campus found that about 30 percent of men and women surveyed did not ask for parental leave for fear of "career implications" (Wilson). "These professors," said the researchers, "seek to hide or minimize family commitments in order to show their dedication to career."

Moving beyond the academic context, a study of medical residents' breastfeeding practices suggests that career pressures play a large role in decisions about breastfeeding for that group (Miller 434–37). While a large percentage of medical residents initiate and maintain breastfeeding during the early period after birth, the study notes a significant drop-off after women return from their maternity leave, usually six weeks. This study indicates the impact of career pressures on women, and the difficulty many women face in choosing breastmilk over formula. One can assume that these women, more than the general population, would have reason to see breastmilk as a better option than formula; even if they received little information on the benefits of breastmilk in medical school, one can assume that their training had provided them with the expertise to conduct their own research.[5] And, indeed, high initiation rates suggest that these women do generally privilege breastfeeding. That the rates had fallen so significantly by the three-month mark, with several women pointing to the pressures of residency as the deterring factor, means that the logistics of managing careers and the need to not appear to be allowing parenting to take precedence over career play a large role in the way many women frame breastfeeding. Several other studies considering the training received by residents specializing in pediatrics suggest that pediatricians "lack knowledge and training on breastfeeding topics," thus indicating that the issue is complex ("Physicians" 587). Shanler's survey of 1,137 active fellows of the American Academy of Pediatrics (AAP) found that only 65 percent recommended exclusive breastfeeding for the first month after birth, only 37 percent recommended breastfeeding at all for the first year, and that 72 percent were unfamiliar with the Baby Friendly Hospital Initiative (BFHI), a program to promote breastfeeding in hospitals. While this initiative involves cooperation between UNICEF and governmental agencies working to promote breastfeeding in other countries, it has met obstacles in the United States primarily because of "the reliance on free formula and other formula company products and gifts that are accepted by many hospitals" (Schanler; "Physicians" 588).[6] Although Schanler's study concludes that "pediatricians have significant educational needs in the area of breastfeeding management," Freed finds that residency training is not adequately preparing pediatricians for their role in breastfeeding support and promotion (Freed; "Physicians" 586).

RESULTS OF THE LACK OF FEMINIST
ATTENTION TO INFANT FEEDING

As a few critics have noted, one result of the lack of feminist attention to infant feeding is that lactating women and the culture at large do not have a feminist framework informing their thinking about child nourishment and breastmilk (Hausman; Van Esterick "Breeding").[7] Without strong feminist advocates, lactation will continue to be defined as an issue pertaining only to women and only for a narrow window of time. Our examination of the environmental contamination of breastmilk suggests the danger of such a perspective. If feminists were more forcefully to embrace this issue, they would help disseminate information so that men and women would be better able to make choices and advocate for changes that affect infant feeding. For example, feminists continue to be leaders in the struggle to break the hold that advertisers, the media, the music industry, and the culture at large have on women's bodies. They are well positioned to advocate for the lactating breast as a legitimate cultural icon. Another result of the lack of feminist attention to the issue has been that we do not have a strong coalition of voices arguing for policies and legislation that would benefit parents and children—from provisions for adequate family leave time to those allowing women to nurse or use breast pumps in the workplace, to subsidies for breast pumps, lactation specialists, and childcare, to greater flexibility for those participating in federal WIC (Women, Infants, Children) programs, to policies assuring that childcare workers are valued, given adequate pay with benefits, offered strong parental leave policies to nurse their own children, and granted opportunities to continue their education. Furthermore, to remain silent on this issue is to go with the default position—to embrace formula, thus colluding with traditional medical models, big industry, and cultural perceptions that regard nursing as better left in the closet. It is to collude with practices that hurt poor women and children in "developing" and "developed" nations, indeed that hurt all of us when we take into account the full impact of allowing the status quo to continue unchecked.

Of the feminist attention to infant feeding that is out there, most does not address, or merely mentions, the problem of the environmental contamination of breastmilk or the ecological effects of choosing formula. We can look at the fact that Pam Carter and Bernice Hausman write eloquently about infant feeding issues and their relation to feminism, yet do not broach environmental contaminants or address the work of such writers as Sandra Steingraber, whose books, while not offering a self-consciously feminist stance, do provide discussions pertinent to feminist discourses on this issue. What's more, Steingraber's eloquent and straightforward approach to scientific studies about the the "breast versus bottle" controversy and the problem of breastmilk contamination is an answer to many of the issues that Hausman and the feminist scholars she negotiates between view as problems and on which they base their sometimes contentious engagement. In this context, it is significant that Steingraber published her second book—*Having Faith: An Ecologist's Journey to Motherhood*—just after Hausman's came on the market, so that Hausman most probably did not know of the material covered there.

I can imagine that Hausman, Carter, and others have embraced with pleasure new attention to breatmilk and have much to contribute to current discussions. To fail to position infant feeding as a significant feminist concern is one thing; but not to position the environmental contamination of breastmilk and formula and the costs of formula-use as significant feminist concerns is to leave unaddressed concerns that affect all women, especially the most vulnerable. This omission means that these more vulnerable women often lack support or do not have access to information about other women in similar situations with whom they might share stories, knowledge, and tactics. Linda Layne provides an example of how a lack of knowledge of one another by women experiencing pregnancy loss due to environmental contamination worked to keep these women isolated; her work demonstrates how coming to see such events as involving structural inequalities can help to bring women together ("In Search" 25–50). In addition, this omission leaves the feminist movement—loosely defined—open to criticism that feminists have worked hard to counter for several years—that it is biased in favor of white, middle-class, privileged, first-world women's concerns. To embrace the environmental contamination of breastmilk as a significant feminist issue would be to foreground an important matter that would do much to draw "feminists" into coalition with those advocating against environmental racism and other forms of environmental injustice; it would foreground intersectionality, which lies at the heart of current feminist praxis.

FEMINIST ATTENTION TO INFANT FEEDING: "THE SCIENTIFIC" VERSUS "THE POLITICAL"?

In one of the most recent and comprehensive considerations from a feminist perspective, *Mother's Milk: Breastfeeding Controversies in American Culture* (2003), Bernice Hausman points out that feminist attention to breastfeeding tends to follow disciplinary lines. She notes that historians Rima Apple and Janet Golden approach breastfeeding "in relation to other social transformations in the United States and Europe," that anthropologists Vanessa Maher and Katherine Dettwyler look at the "effect of cultural ideals, material practices, and economic structures," and that literary critics Mary Jacobus and Marilyn Yalom investigate "lactational symbolism" and representations of breastfeeding (196). Such richness suggests the power of breastfeeding as an image, an icon, and a relationship that reaches deep into several aspects of our lives. And it reminds us that feminist studies and women's studies, as truly interdisciplinary fields, occupy positions well situated to embrace the environmental contamination of breastmilk.

Hausman divides feminist scholars into two camps, and argues that we can see them as either privileging "scientific evidence" (thus advocating for women to use breastmilk instead of formula) or opting for "political readings" that downplay scientific studies, thus arguing that women gain more in being free to choose formula. We can look to the work of Jules Law, Linda Blum, and Pam Carter as three examples of those critics who refute scientific evidence to varying degrees. Jules Law most clearly articulates a position in favor of the "political" interpretation. Law believes that the scientific arguments for breastfeeding make assumptions about the traditional sexual division of labor and are "highly polemical"; indeed, for him, what gets presented as scientific evidence is "misleading at best and false at worst," and often "not based on medical research at all" (412, 407). He claims that such studies "foreground one dimension of the issue (risk) disproportionately while relegating another (the sexual division of labor) to an almost invisible background" (408). Infant feeding, for him, should be conceptualized as being about "balancing the labors, pleasures, well-being, development, and opportunities of a household's various members," not about addressing what he sees to be minute advantages for breastfeeding babies (426). Consideration of Carter's and Blum's arguments suggests that they, like Law, see the physiological benefits of breastfeeding—at least in industrialized countries—as more or less limited. Breastfeeding advocacy, for them, must be viewed as a political act, an attempt, at least at some level, to promote "intensive mothering."

Law states that "the most dramatic risks of formula feeding are generally based on universalizing inferences from situation-specific dangers in hygiene-poor, poverty-stricken, nonindustrialized settings" (410). His complaint is that "breastfeeding advocates continue to allege substantial mortality rates associated with formula feeding even in Western industrialized settings" (410). For Law, therefore, breastfeeding becomes synonymous with obstacles "preventing work outside the home" (408). Bearing in mind how far women still have to go—they earn still only 75 cents to every dollar a man does and occupy few positions at the top of governments or corporations—Law and those in his camp believe breastfeeding advocates ask women to sacrifice or impinge on their careers and their autonomy. Again, they dismiss pro-breastfeeding arguments as based on faulty studies or on the transference of findings regarding the developing world onto first-world contexts. Law's reading allows him to claim that "only a culture that assumes the greatest social good is obtained by a particular half of its members dedicated to one-on-one child care could weigh marginal fluctuations in normal childhood infections against the life-world of half of its citizens" (426).

FEMINISTS CRITIQUING (AND EMBRACING) SCIENCE

Because science, like other fields of knowledge that have been seen as objective and not socially situated, has been used to devalue women and keep them in subordinate roles, it makes sense that feminists continue to be cautious about valuing scientific studies without at the same time questioning their social embeddedness. We should be careful of situations in which science gets used to further the status quo without being seen as socially situated. Through the years, feminists have lodged powerful critiques of science that have had important results for women (Fox Keller; Harding; Hubbard; Mies and Shiva; Shiva; Spanier). Ruth Hubbard, for example, reminds us that "facts aren't just out there," and urges us to ask "what criteria and mechanisms of selection . . . scientists use in the making of facts?" (119). She points out that "making facts is a social enterprise" (119). Addressing what Law and other feminist critics find objectionable in breastfeeding advocates' arguments, he explains that "an entire range of discriminatory practices is justified by the claim that they follow from the limits that biology places on women's capacity to work" (122).

Indeed, looking at historical examples of how biology has been used to keep women from entering many professions, and noting the bias of such practices as when poor women and women of color have been seen as capable of grueling work by applying a different set of standards, many feminist today consider the methods some science continues to use as mystifying, intimidating, and oppressing women and other nonprivileged groups. One key example of the kinds of contributions feminists have made in critiquing science involves the realization that, until feminists raised the alarm, scientific studies of drugs were conducted on 150-pound males, rather than on women or children whose endocrine and other systems often experience vastly differing effects of various drugs (Boston *Our Bodies* 689). Feminist geographer Joni Seager, who looks more specifically at feminist perspectives on environmental issues, offers another example, noting that "until women started organizing around these issues, the impacts of pollution on women's health were ignored by mainstream environmental organizations, by official health monitoring organizations, and by the biomedical research establishment" ("Rethinking" 1). Indeed, she notes that "questions about women's health and pollution, until recently, were not examined, not taken seriously, and not followed up."

If we look at the specifics surrounding breastfeeding, we see that no one is really disputing the science; the "critics of science" do not deny that breastmilk has been shown to offer more benefits than formula. They differ on how much of an advantage breastmilk offers babies, and how to balance this with the costs to the mother of breastfeeding. In the best-case scenario, policies to lighten the burden on breastfeeding mothers would be in place. Setting up nurseries at workplaces and allowing all women time during the workday to nurse or express breastmilk, subsidizing dual-action, electric breast pumps for all women who wanted them, and mandating the designation of suitable places for women to express milk at work and in the world at large would certainly make breastfeeding easier. Further, ensuring that childcare workers receive good pay, ample benefits, opportunities to further their education, and that their work is considered valuable and legitimate would advance women's rights. Clearly, at this point, some of these arrangements are available to some women, but certainly not to all. This, it seems to me, is where we all need to come together. While it seems unhealthy to be so adamant in our support of breastmilk that we blame women who cannot nurse, it seems unhealthy in other respects to argue that because it is challenging or impossible for some women to nurse at this time, we should refute the science, reject breastmilk as impinging on women's autonomy and endeavors for equal rights, and seek to undermine the fight for making breastmilk an option for as many women as possible.

Furthermore, to assume that for a woman to opt for breastmilk requires her to give up her work outside the home is to ignore the reality of many women who are able to continue careers while nursing young children. I have been able to continue my work as a professor while giving my baby nothing but breastmilk until he was seven months old. However, I am one of the lucky ones. I have a job that allows me to easily express milk at work; I have childcare that is subsidized by my union and so is able to provide benefits to its childcare workers; I have access to the Internet so that I was able to purchase an affordable (via e-bay) dual-action breast pump that allows me to express up to eight ounces of milk in seven minutes, and I have been fortunate to have lived in areas relatively safe from environmental toxins. Clearly, to assume that all women can make such choices is to ignore the experiences of many.

It is important to bear in mind the critics of science's (Hausman's term) insistence on approaching the topic through "attention to women's subordination," and their desire for "some acknowledgment of infant formula as a healthful option for feeding babies." I think it is important to decouple these two concerns (211). If we look at problems with using breastmilk while resuming paid work (such as lack of logistical support), we see that similar problems are associated with engaging in paid work while using formula. Parents in such situations confront some different issues—for example, the high cost of formula as opposed to the high, but still much lower expense of a breast pump—but these issues illustrate a similar structural situation involving sexism, classism, racism, and other forms of oppression. For example, whether one uses formula or breastmilk, someone else, most often someone more oppressed, will be employed to make the situation work, often without benefits. Looking at the issue through the lens of "women's insubordination" is key, yet this must involve taking into account those who frequently get ignored in such discussions.

On the other hand, the "critics of science" feminist scholars who regard formula as a "healthful option" offer a corrective to the position put forth by many breastfeeding advocates that women are to blame for not nursing—including women who cannot use breastmilk for a range of reasons. Van Esterick's position that "the trajectory goal becomes not to have every woman breastfeed her infant, but to create conditions in individual households, countries, and nations so that every woman could" offers what seems to me to be a valuable perspective (Beyond 211). What needs to be added here is the reminder that even if conditions were to support using breastmilk, there would still be instances when women cannot—and these women do not deserve to be put on a guilt trip. Institutional support is the key factor.

If we return to Hausman's configuration of the "scientific" versus the "political," we find that a key example of an approach that embraces the "scientific" is the work of Katherine Dettwyler ("Beauty"; "Promoting"). Dettwyler's anthropological work considers scientific studies in order to argue for the benefits of breastfeeding, while at the same time asserting the importance of social context in women's lives. For example, she considers tobacco smoking and how scientific studies have played a role in changing attitudes about its health effects and policies surrounding its use ("Beauty"). This provides an important model for breastfeeding. However critics try to position it, the science points, loud and clear, to the advantages of breastmilk over formula. With tobacco, once the science became unequivocal, the legislation followed. For example, just a few years ago many would be surprised that today in many locations people can no longer smoke in public spaces. We could take this as a model for how we could use science to launch changes in attitudes, legislation, and behavior surrounding breastfeeding (210).

Having pointed to "the science" and its establishment of breastmilk as offering more benefits than formula, Dettwyler goes on to remind us that one of the greatest contributions feminists have to offer on this issue involves theoretical approaches to the role of the breasts in patriarchal cultures. Foregrounding this issue suggests that choices about how to feed one's infant have to do, not just with what I call structural logistical issues (e.g., the kinds of arrangements or lack thereof one finds to support using formula or breastmilk), but also with what I call structural/social issues (such as how one feels about the social implications of implementing one choice or the other). I have spoken with several friends who could, logistically, have used breastmilk and returned to work, but chose not to because, in one woman's words, "I felt like a cow," and in another, "I could tell my husband didn't find me attractive [when I nursed]." Others complain that "nursing hurt" or that pumping "was just too much of a pain." So, in other words, to refute the "science" and embrace formula as a comparable choice is to skirt the whole question of the role of perceptions about nursing within specific cultural contexts. Dettwyler critiques the assumption that the primary purpose of women's breasts is for sex (i.e., for adult men), not for feeding children, thus offering commentary that brings feminists together.

Penny Van Esterick, an anthropologist who has written on breastfeeding and feminism, but has been largely ignored by the few feminists who do address infant feeding, offers some important reminders about how we might resituate breastfeeding in a capitalistic setting.[8] She points out: "If the work of lactation is valued as productive work, not the duty of a housewife, then conditions for its successful integration with other activities must be arranged" (*Beyond* 75). In addition, she cautions us against separating so-called first- and third-world contexts, as many feminist critics [those who dismiss scientific claims of breastmilk's superiority to formula] have done in the name of downplaying the health advantages of breastfeeding in developed countries (195). While conditions in developing countries may make formula-use more detrimental to infant and maternal health in those contexts, to argue that the benefits conferred by breastfeeding in first-world contexts are marginal in comparison—and thus should not enter into our discussions of infant feeding—is faulty reasoning.

Such a perspective divides women from developed and undeveloped countries, as well as wealthier and more disadvantaged women within these regions (*Beyond*). Furthermore, it ignores the fact that formula-use on a large scale comes with other problems that affect all of us—not just babies and infants. It bears repeating that this comes with ecological costs and allows us to skirt the problem of the environmental contamination of breastmilk. In addition, scientific studies have unequivocally pointed to the benefits of breastmilk for babies in all regions. In response to fears that breastfeeding has been associated not just with models of intensive mothering and politics keeping women in the home, but also with politics advocating traditional family structures, it is important to note that adoptive moms can nurse, and that using breastmilk can work for less traditional family structures, including lesbian households in which both moms could nurse or express milk, and gay households where banked milk could be used. Although some of these arrangements clearly require more logistical maneuvering than others—for example, an adoptive mom would benefit from access to a good breast pump and would have to devote some time to developing a milk flow—they do remind us that using breastmilk need not accompany the model of intensive mothering associated with its nuclear family and its sexual division of labor.[9] The use of "science" to promote breastmilk need not be associated with regressive politics.

BEHIND THE "POLITICAL" ARGUMENT: FORMULA AS A LIBERATORY TECHNOLOGY

While Law and others in his camp situate their arguments as having to do with how we position science, a look at Shulamith Firestone's work suggests that Law's argument represents one in a tradition of feminist debates about the sexual division of labor. Law suggests that his preference for formula over breastmilk is the result of his reluctance to use "science" to burden women further by imposing more dictates about how we should live our lives. But consideration of an important feminist discussion of "social/reproductive technologies" and how these have been positioned, historically, in relation to the women's movement and its fight for "equal rights" suggests that there is a precedent in feminist thinking for linking breastfeeding with the sexual division of labor. Writing in the early 1970s, Shulamith Firestone, an important figure in second-wave feminisms, considered the idea that sex discrimination stems from the "sexual division of labor," and the idea that the social/ reproductive technologies of birth control and baby milk formula offer freedom (183–86). In this context, I am reminded of meeting a woman who proclaimed, upon hearing that I teach Women's Studies, "I'm not a feminist because I don't think men and women are equal." The slippage here from demanding equal rights to "being equal" to "being the same" shows how the marks of the "biological" and the "material" continue to plague feminisms, but also that feminists have much public relations work to do. In *The Dialectic of Sex,* Firestone illustrates how cultural recognition of biological differentiation leads to sex-based discrimination. She notes that sometimes we see gender differences as "a superficial inequality," one that can be "solved by merely a few reforms, or perhaps by the full integration of women into the labor force" (183). "But," she continues, "the reaction of the common man, woman, and child 'That? You can't change that!' . . . is the closest to the truth." Claiming that "this gut reaction—the assumption that, even when they don't know it, feminists are talking about changing a fundamental biological condition—is an honest one," she goes on to promote a "feminist revolution" that solves, for her, the problem of sex-based discrimination.

Targeting reproduction and childcare, Firestone points out that since women throughout history "were at the continual mercy of their biology," and since the "basic mother/child interdependency has existed in some form in every society," the "natural reproductive difference between the sexes led directly to the first division of labor based on sex." For Firestone, "the elimination of sexual classes requires the revolt of the underclass (women) and the seizure of control of reproduction: the restoration of women of ownership of their bodies, as well as feminine control of human fertility, including both the new technology (fertility control) and all the social institutions of childbearing and childrearing" (183). Here, access to reproductive technologies, freedom from various incursions on the body, as well as freedom from breastfeeding and exclusive childcare are necessary conditions of women's autonomy. Positing that "the end goal of feminist revolution is not about just breaking down the sexual division of labor," and not just about "the elimination of male privilege, but the elimination of the "sex distinction itself," Firestone introduces problems later to emerge in deconstructive and constructivist arguments and marked by the publication of Judith Butler's *Gender Trouble* and *Bodies That Matter*—among other books—in the early to mid-1990s. The bottom line, for Firestone, is that "the dependence of the child on the mother (and vice versa) would give way to a greatly shortened dependence on a small group of others in general . . . the tyranny of the biological family would be broken" (186)

For Firestone, as for other feminists, childbearing and childrearing stand as impediments to overcoming patriarchal systems of governance. While pumping and storing one's milk during the workday (so that someone else could feed the baby) would be prioritized over nursing in the domestic sphere, neither seems to constitute full "freedom" for Firestone and those in her camp. In this discussion, maternal breastfeeding, positioned as an integral part of the sexual division of labor, gets constructed as something that must be not only deconstructed, but eliminated. Thus, formula gets privileged over using breastmilk. Jelliffe and Jelliffe, two historians of breastfeeding, point to a marker of the degree to which breastfeeding has historically been portrayed as out of sync with the liberated woman when they note: "As with cigarette smoking, bobbing the hair, and the contraceptive diaphram, the feeding bottle was often visualized by the 'flapper' of the 1920s as a symbol of liberation and freedom" (189). And Rima Apple, in pointing to the popular 1930s advertisement for Pet Milk, which heralded freedom for mothers with the slogan "Take the baby and go!," suggests the popularity of such a connection. She addresses this notion in her study *Mothers and Medicine: A Social History of Infant Feeding*, noting that "allegedly, increasing numbers of women refused to breastfeed because nursing tied them down," as "changes in American society expanded activities outside the home," and as "widening educational, organizational, and occupational opportunities, which resulted from women's social and political action, drew women outside the home" (173; 182). Indeed, other scholars have made similar observations of a later generation, pointing out that in the 1960s and 1970s people portrayed formula as "liberation in a can" and as a tool allowing women to leave behind domestic duties for public geographies and market economies (Galtry; Jelliffe and Jelliffe). What is telling is that none of these discussions addresses the fact that if the mother is being "liberated," someone else more oppressed socially is being "tied down."

WHO'S DOING THE FEEDING? RACISM, CLASSISM, AND COLONIZATION IN CHILDCARE WORK

Whether one is concerned with formula-use or with breastmilk, if we are talking about women's roles in the social machinery and the sexual division of labor—women returning to work, beginning a career, going back to school, or taking an afternoon off from a career as a full-time mother and "house-wife"—if a woman has any autonomy from her child, someone has to do the feeding. Increasingly, fathers, same-sex or opposite sex domestic partners, relatives, or friends are asuming this role. But the reality is that, for the most part, those who do childcare work tend to be women, and women made more vulnerable by systemic oppressions continuing to operate in our culture— women of color, women of lower economic standing, immigrant women, single mothers, or women in their teens or early twenties or at or beyond retirement age. We often forget that while middle- and upper-class white women talk about their freedom to pursue careers or to be stay-at-home moms, many other women do not experience this freedom, and have not, historically, enjoyed such choices.

Another related issue concerns the assumption that formula offers "liberation." In *Mothers and Medicine: A Social History of Infant Feeding*, Rima Apple suggests that although formula has been marketed as offering "freedom"—as in the 1902 article in the popular journal *Babyhood* that claimed "supplemental feeding" would give mothers "a little freedom," and the 1924 Pet Milk ad previously mentioned encouraging mothers to "take the baby and go!"—its increasing adoption from the early 1900s to the 1960s in the United States had as much to do with self-interested marketing practices, the rise of "scientific motherhood," and the squelching of women's confidence in their ability to provide adequate milk, as it did with their perception that the "new technology" offered freedom to pursue careers "outside the home" (109, 158). What interests me is the degree to which such factors continue to influence choices about infant feeding. Beliefs that formula is associated with liberation from domesticity do not hold up. In sketching a genealogy of public school home economics courses from the 1920s to the 1950s throughout the United States, Apple notes that "educators feared that the feminist movement with its insistence on equal rights for men and women taught girls to consider career first and motherhood second, if at all" (115). To counter this, they sought to "make motherhood women's profession." But breastfeeding did not appear on the job description. Here the prescription for women occupying domestic spaces, and playing the role of the "good mother," involved using formula. And the inculcation started early; breastfeeding was not part of the picture presented to the Little Mother's Clubs of the period. They promoted "scientific childcare" with its insistence on formula-use. In addition, courses offered by state health departments and education agencies reached out into the community with such classes as "The Care of Infants" touting formula-use. To provide a feel for the popularity of such classes, it is useful to note that more than 45,000 women had taken the Indiana State Department of Health's course by 1928 (Apple 116).

Several studies imply that the assumed relationship between working outside the home and using formula is complicated (Apple and Golden; Galtry). For example, between 1940 and 1960, fewer mothers were working outside the home than at other times yet this period was the zenith of formula-use in the United States (Apple 150–68). In these cases, the assumed link between "exclusive mothering" and breastfeeding complicates the discursive framing of breastfeeding as linked to women choosing to forego careers in order to engage in full-time childcare. Current evidence indicates that while more new mothers than ever are participating in the workforce in the United States, breastfeeding rates here have gone up over the last decade, thus suggesting that such participation does not necessarily correlate with formula-use.[10] On the other hand, evidence also suggests that certain kinds of paid work discourage nursing or pumping, and that jobs that allow pumping are advantageous because they include the option for mothers to have someone else feed their babies. Women who tend to hold higher-paying, "career" jobs—who usually are college-educated, white, and older—are also the mothers who tend to initiate nursing and continue nursing or using breastmilk longer ("An Easy"; Appea; Carter).[11] These are also the women who enjoy greater flexibility around workplace policies as well as more options to pump or nurse while at work (Galtry). For these women, formula is less associated with the "freedom" to work outside the home than it is for other women.

The "technology" of formula sets women "free" to work outside the home in those cases in which opportunities are not available for them to have time off to establish breastfeeding or to take breaks to pump or nurse at work. These women tend to be less economically advantaged, less educated, younger, and women of color—the women who, not surprisingly, tend to breastfeed at lower rates (Appea; "Child Health"; Wambach and Cole 282–94). For example, a 1998 survey, "The Health and Human Services Blueprint for Action on Breastfeeding," reports that while only 29 percent of all mothers in the United States were still nursing six months after giving birth, the figure for African American mothers was even lower, at 19 percent (Appea).[12] While "white," middle-upper-class, heterosexual, and educated men and women—the privileged norm—carry around what Peggy McIntosh has called "an invisible backpack" that functions to help them to procure higher paying jobs with greater flexibility around childcare and feeding, women of color and younger women with more limited economic means tend to occupy positions with fewer options; they also become childcare workers who allow other women to work outside the home and nurse young children (95–105).

While many working parents successfully incorporate breastmilk into their feeding practices, many women, including those able to take time off from work to give birth, initiate breastfeeding, and express milk, do turn to formula when they return to the workplace or soon after. This is particularly true of certain kinds of work. For example, the previously mentioned study of breastfeeding practices among resident physicians found that while 80 percent of sixty residents initiated breastfeeding and continued throughout their maternity leave (with a mean of seven weeks), the breastfeeding rate dropped to 15 percent at six weeks (Miller, Miller and Chism). Of those who continued to use breastmilk, 79 percent felt that their work schedules offered insufficient time to pump at work and only 54 percent felt supported by their attending physicians. This points not only to the challenge for women to continue using breastmilk while working (however prestigious the job), but also demonstrates that the recommendations to breastfeed for at least one year from several representative medical bodies (the American Association of Pediatrics, the Surgeon General, the board and editors of *JAMA*) do not fully correspond with the actions of some medical practitioners.

In addition, many more women who work in settings that should be covered by the scanty legislation that does exist to protect mothers' right to pumping breaks and spaces other than bathrooms in which to pump do not find management accommodating, and have to push for arrangements if they want to express milk. So the question arises: what circumstances do other women encounter—those who clean the houses, run the errands, and feed and diaper the children of those mothers who chose to resume paid work? Do they have opportunities to take advantage of pumping technologies? Do they use formula? Is it subsidized? Do they get paid leave to care for their children when they are sick, or subsidies for breast pumps, and time off for nursing breaks? As many scholars have pointed out, childcare work is one of the lowest paying jobs, with fewer benefits than most. According to a study conducted by Smart Start, 42 percent of childcare centers do not offer a single day of sick leave, despite workers' close interaction with sometimes sick children; only 16 percent offer full health insurance. I expect that the rates are more variable, but lower for those who work in the home, especially for healthcare benefits, which seem to be nonexistent for most in-home workers (*LHMU*). This is a woman's issue, and a feminist one. The Smart Start study also shows that 99 percent of these workers are women; 71 percent have children of their own and 22 percent are single mothers; not surprisingly, the number receiving public assistance grew in just three years, from 26 percent to 34 percent (*LHMU*).

IMMIGRANT WOMEN DOING CHILDCARE

Shellee Colen, studying what she calls stratified reproduction, which designates ways in which "physical and social reproductive tasks" get accomplished differently according to inequalities based on "hierarchies of class, race, ethnicity, gender, place in a global economy, and immigration status," looks at West Indian childcare workers and their employers in New York City (78). She notes that "as upper-middle and upper-class women find higher end positions, more women, especially women of color, including new immigrants, find positions on lower rungs" (81). Pointing out that workers often encounter a stratification, ranked according to their status as illegal immigrants, green-card holders, and citizens, undocumented live-in workers often experience "indenture-like conditions" that "they often compare. . . to slavery" (79–80). Low wages, no medical or retirement benefits, long hours (since work must extend on either side of the "normal" workday) also include long absences from the workers' own children. While I have not been able to find information on the percentages of childcare workers who use breastmilk or formula, it seems unlikely that many of these women have opportunities to choose between infant feeding options. For example, language barriers often make arrangements for pumping breaks challenging, while the lack of pumps in many regions outside the United States means that many immigrant workers from some regions might not have heard of breast pumps. Studies have shown that for immigrants coming from regions with high breastfeeding rates, those rates tend to drop significantly in relation to the amount of time spent in the host country, particularly the United States ("Breastfeeding Practices in Los Angeles" 5; de Bocanegra; Romero-Gwynn and Carias 626–32; John and Martorell 868–74). And because of lack of state-sponsored childcare, the ideology of privatized childcare, the lack of options for parents, and the unwillingness of many native-born workers to take such jobs, "[t]his system of reproduction assigns paid reproductive labor to working-class women of color" (2). At the same time, "the system itself reproduces such stratification."

While some domestic and childcare workers across the globe have begun to unionize and demand and procure better working conditions, the link between the exploitative arrangements prevalent in childcare and the effects of globalization suggest that unless people begin actively to address these issues, the situation will only get worse, mirroring the growing gap between the "privileged" and the "nonprivileged" within the United States and globally. Commenting on this, Colen suggests that this gap is "framed by a transnational stratification system" (2). For example, confronting "the legacies of slavery, colonialism, underdevelopment, and Carribean articulation into a world capitalist system and the constraints these place on fulfilling their gender-defined obligations," many West Indian childcare workers toil under a global system that has been shown to be a result of "the direct relationship between worsening economic conditions, including those induced by the International Monetary Fund policies" and other global systems of exchange (2). Feminists have much to offer in such situations where individual women find themselves oppressed and yet lack a voice or organizational structure that would help them fight for their rights.

(IN)ATTENTION AMONG FEMINISTS TO THE ENVIRONMENTAL CONTAMINATION OF BREASTMILK

It makes some sense that because the feminist perspective on breastfeeding is so ambivalent, we would not observe much feminist attention to the environmental contamination of breastmilk. But, on the other hand, when we look at some of the claims ecofeminists and other ecologically minded feminists have made about why environmental issues are or should be considered feminist issues, this inattention seems surprising. As Joni Seager points out, "feminists have been particularly active in reframing the ways in which environmental relations in general and pollution in particular are understood" ("Rethinking" 1). In addition, feminists have been powerfully vocal about issues of violence against women and children, the curtailing of women's reproductive rights, and racism, classism, ethnicity, colonization, and other oppressive practices. The environmental contamination of breastmilk is all of these and more. Indeed, Barbara Smith's definition of a feminist praxis, representative of many feminists' understanding of a feminism they support, is one that challenges all forms of social and economic discrimination and oppression: "Feminism is the political theory and practice that struggles to free all women, women of color, working-class women, poor women, disabled women, lesbians, old women—as well as white, economically privileged, heterosexual women" (49).

Karen Warren, articulating an ecological feminist perspective, points out that "important connections exist between the treatment of women, people of color, and the underclass on the one hand and the treatment of nonhuman nature on the other" (3). For her, supremacist, dominating, and oppressive thinking and actions — whether the oppressed are women, people of color, the poor, disabled people, children, nonhuman species, or various ecosystems—all work together to systematically limit the rights of the less privileged. Most feminists would agree that we must struggle against such oppression and domination in all its forms. In the words of feminist scholar Ynestra King, ecological feminism's "challenge of social domination extends beyond sex to social domination of all kinds, because the domination of sex, race, and class and the domination of nature are mutually reinforcing" (20).

In her first book, *Living Downstream: A Scientist's Personal Investigation of Cancer and the Environment*, Sandra Steingraber who has a PhD in biology, questions a system that allows those in power to make decisions that affect us all. She points out that "many . . . ask how we can claim liberty when our own bodies—as well as those of our children—have become depositories for harmful chemicals that others, without our explicit consent, have introduced into the air, food, water, and soil" (xvi). Intent on "break[ing] the silence that has so long surrounded the topic of cancer's ecological roots," she makes it her mission to help people become aware that hegemonic forces work to undermine our health and well-being, and strives to bring people to rally against such injustice (xvi). Steingraber truly is one of "Rachel's daughters"—a biologist, ecologist, a cancer survivor, and a writer, one who, like Carson, writes with eloquence, beauty, and a straightforwardness coupled with passion about the effects of the current environmental crisis. She may not be identified by others as primarily a feminist writer,[13] but her clear discussions of the environmental contamination of our rivers, air, soil, animals, and people are forceful and indict those in power who allow such practices to continue, thus putting her work in line with other ecologically minded feminists. Steingraber's work on the benefits of breastfeeding, and the environmental contamination of breastmilk in her second book, *Having Faith*, offers much that can help us sort through some of the controversies in feminist discussions of infant feeding.

On the other hand, Jules Law's work reveals a certain frustration with feminists who appear to embrace science uncritically. There seems to be a belief that science does not have all the answers, and doesn't take social and political factors into account. While we must continue to be wary of uncritically accepting the findings of anything calling itself a scientific study, it is possible to assess the relative validity of various studies. We can look to literature reviews and other critiques of groups of studies to bring us to an acceptable confidence level about certain issues. Law complains that "the strictly medical benefits (to both mother and child) of breastfeeding in the early weeks and even days after birth receive disproportionate attention, although one virtually has to read between the lines of the extant literature to see this" (426). She continues: "To a working mother, for instance, it might be more important to know whether the bulk of breast milk's immu-nological benefits are passed on to her baby by two, four, or six weeks of breastfeeding (since these differences bear on tangible maternity-leave poli-cies and demands) than to know that a year of it reduces by a fraction of a percentage point her baby's chance of getting a rare childhood disease" (426). Steingraber's book, which came out one year after Law's 2000 article, helps to answer some of those questions, particularly the valid inquiry Law has about the relative benefits of breastfeeding for various lengths of time.

WHAT SCIENCE CAN OFFER: THE BENEFITS OF BREASTMILK

On the question of the benefits of breastmilk in comparison to formula Steingraber speaks in the vein of Carson, conceding the difficulty of coming to consensus. She notes: "Studies documenting the benefits of breastfeeding on infant health are not as easy to undertake as they might at first seem . . . Breastfeeding mothers in the United States tend to have more money and education than mothers who bottle-feed. They also smoke less" (*Having* 226). But, she points out, "The best studies take these social factors into account and correct for them." She continues to describe "one of the most convincing studies," published in 1998, that considered 2,000 Navajo infants before and after a breastfeeding promotion program. Noting that 16 percent of the infants were nursed before the study, and 55 percent after, the study found that infant pneumonia fell by a third and gastrointestinal infection by 15 percent. Its designers concluded: "Increased incidences of illness among minimally breastfed infants is causally related to lack of breastmilk" (226). In addition, Steingraber finds that "studies consistently show that children and young adults who were breastfed suffer less from a range of conditions including allergies, asthma, type I (juvenile) diabetes, Crohn's disease, ulcer-ative colitis, and juvenile rheumatoid arthritis" (227). Citing several studies, she also points out that "breastmilk may also safeguard against obesity and cancer," and that "these benefits linger" (227).

Here, beginning to answer Law's duration question, she notes a German study indicating that babies breastfed for one year were one-fourth as likely to be obese later in life as those breastfed less than two months (228). A study published in *JAMA* after Steingraber's book appeared corroborates these results, suggesting that babies breastfed for several months are much less likely to be obese as adults—a significant finding given the rising incidence of obesity and related health effects in the United States (Gillman et al. 2461–67). In looking at cancer, "several carefully designed studies have found that artificially fed infants, as well as those who are breastfed only briefly, go on to suffer significantly higher rates of Hodgkin's lymphoma than babies breastfed for six months or more" (*Having* 228). In addition, Steingraber points out that "the fortifying effects of mother's milk on health endures long after weaning," noting a Scottish study that followed 600 breast and bottle-fed babies into school age. Correcting for such factors as economic class, weight, and maternal blood pressure, the study found that the children who had been breastfed had significantly lower blood pressure.

When it comes to questions of benefits on cognitive functioning, motor development, and intelligence, the findings are striking. Steingraber notes that "an impressive battery of studies" indicate connections between brain development and breastmilk. For example, "At age three and a half, children who were breastfed at least six weeks showed more fluency of movement," while "other studies reported that breastfed infants are more mature, secure, and assertive, and score higher on developmental tests" (*Having* 240). Indeed, "They are somewhat less likely to suffer from learning disabilities and have higher IQ s when tested as seven- and eight-year olds"(241). A 2002 study shows that "independent of a wide range of possible confounding factors, a significant positive association between duration of breastfeeding and intelligence was observed in two independent samples of young adults, assessed with different intelligence tests" (Mortensen et al. 2365).[14] Addressing the question that Law and others ask—"are these differences really attributable to breast milk?"—Steingraber notes that "perhaps mothers who breastfeed are simply more likely to possess superior parenting skills or value education more highly or have more money and time to spend on brain-enhancing activities for their children" (241). Conceding that "at least one study found that differences in IQ scores disappeared once socioeconomic factors were corrected for," she also points out that "in most studies, significant differences remain" (241). And the 2002 study published after her book went to press bears this out.

In addition, Steingraber consolidates findings on the advantages of breastfeeding for the mother and concludes that breastfeeding benefits mothers in several respects. For example, breastfeeding makes a nursing mother produce the hormone oxytocin, which works to return the uterus to its prebirth size more quickly. Furthermore, a little known fact that I can attest to from my own nursing experience—and one that would be experienced by adoptive moms—is that when the body produces oxytocin, dubbed "the love hormone," it causes us to feel much more calm and happy, and sends a sense of well-being throughout our systems. Apparently, this is nature's way of helping women to cope with the stresses that can often accompany having a new baby. One friend confided that when she stopped nursing she felt much more short tempered with her two-year old and much less able to calmly administer lessons about how we, in the adult world, handle certain problems that two year olds tend to scream about. When I had my own child, I understood what she meant. Other benefits to nursing mothers include lower rates of ovarian and premenopausal breast cancer; Steingraber cites an Icelandic study that found that breastfeeding significantly reduces the chances of contracting breast cancer before the age of forty, and that the longer the duration of breastfeeding, "the greater the protection afforded" (228).

WHEN THE "PERSONAL IS THE POLITICAL": THE ECOLOGICAL IMPACTS OF FORMULA-USE

While it is certainly important to consider the effects of breastfeeding and formula on infants, children, and nursing women, it is also important to view those effects on a larger scale. In other words, we should consider the ecological impacts of formula production and use. Before turning to specifics, we must bear in mind that whenever we examine the environmental costs of certain behaviors and practices, we must also consider that ecological costs will be borne more fully by the most vulnerable populations—people in developing nations and the less privileged in all nations. Formula production creates pollution, and the development and discarding of packaging involve more waste, including dioxin used in labeling and plastics in containers. The finished product must be transported—again, an activity that produces waste. Radford, who writes about the aggregate of ecological impacts of formula-use, draws an analogy between the idea that breastmilk should be replaced by artificial substitutes and the concept that kidneys should be replaced by dialysis machines. For him, while both "have a role to play and can save lives," to use them in place of the simpler alternative is "a waste of resources" (1). Writing in 1991, he offers a wealth of statistics that though dated, still indicate the contours of his approach. For example, quoting 1978 figures from UNICEF, he notes that "if every baby in the USA is bottle-fed, almost 86,000 tons of tin plate are used up in the required 550 million discarded babymilk tins" and adds that paper labels constitute "another 1,230 tons of paper" besides that used for promotion (2). Clearly, packaging has changed. But the waste and pollution associated with the production of formula add up when we consider the amount of formula required to feed one child.

Without going into the specifics of his analysis, some of the issues he addresses require us to consider that the production of soy- and milk-based formulas involves land for grazing or growing, water for production, and other resources for transportation and use, the fact that most lactating women do not menstruate and therefore require no tampons or pads, thus cutting resource use and the production of waste, and that breastfeeding lowers fertility rates and prevents more births than all other forms of birth control combined, thus reducing the strain on earth's resources. Because Radford offers this insight without an examination of the sometimes oppressive politics involved in discussing population issues—for example, that advocates of population "control" often blame poor women, women of color, and women from less developed nations rather than the consumption habits of the first world—realize the degree to which we need a feminist analysis informing this discussion. Van Esterick, while also addressing some of these issues, begins her discussion from an angle more clearly identified as feminist; she notes that "breastfeeding challenges the predominant model of women as consumers" (*Beyond* 72).[15] As Vandana Shiva, surveying the increasing burdens on earth's carrying capacity and the increasing destruction of its ecosystems, points out: "If we want to reverse that decline, the creation, not the destruction of life must be seen as the truly human task, and the essence of being human has to be seen in our capacity to recognize, respect, and protect the right to life of all the world's multifarious species" ("Impoverishment" 88).

FINDINGS ON THE ENVIRONMENTAL CONTAMINATION OF BREASTMILK AND RESULTING HEALTH EFFECTS ON WOMEN AND CHILDREN

In confronting what science can offer on the question of breastmilk toxicity, we find a well-documented account. As Rachel Carson pointed out so long ago when she wrote *Silent Spring*, the first report of the encroachment of toxins in human breastmilk came out in 1951 in a paper called "Occurrence of DDT in Human Fat and Milk" that considered the presence of DDT in the milk of women of African American descent living in Washington, DC (Laug, Kunze, and Prickett 245–46). Sandra Steingraber comments that her "office shelves contain stacks and stacks of published reports documenting the presence of environmental chemicals in human milk"; I can say the same (*Having* 253). I can also say that thanks to the work of Steingraber and others, as I write this I am also quite literally nursing my own infant son, as Steinbraber nursed her daughter Faith while she worked on her manuscript. We do this even though we both know that the milk of all mothers is contaminated with various toxins that have been dispersed throughout our environment.

If Steingraber, who writes with such detail and clarity about how flame retardants (used in computers and other technologies), dry-cleaning fluids, fungicides, toilet deodorizers, cable-insulating materials, and chemical by-products of garbage incinerators, among others, invade our breastmilk, and does so while nursing, I can trust that I am making the better choice in deciding to nurse.

But I know as well as she that such a choice is more agonizing for mothers who live in geographies that make them more susceptible to environmental contaminants, and that unless we collectively act to change these circumstances, our daughters and their daughters will face much more difficult choices. On the question of the effects of such contaminants, Steingraber, never willing to sensationalize or take the easy route, points out, "It is one thing to document the presence of contaminants in breastmilk. It is another to document evidence of harm" (*Having* 268). Why? One answer is that to document this, we would have to have a "control group"; we would have to compare the breastmilk of a group of women with contaminated milk with that of women whose breastmilk is untainted. But no woman's milk would qualify for such a group; all of our milk shows signs of environmental toxicity. Another problem with conducting such studies involves exposure to pollutants that infants get while in the womb, and it is difficult to sort out when various contaminants have invaded the body. If we look at animal studies, however, the findings suggest pronounced effects. According to Steingraber, "Animal studies . . . consistently show that POP-contaminated breast milk contributes to structural, functional, and behavioral problems in offspring" (269). Such effects have been shown to take place at levels comparable to human exposures currently seen in certain populations (269). Monkeys, for example, exposed during infancy to PCBs at levels found in human milk demonstrated "a decreased ability to learn and master new tasks" compared to controls (269). In fact, the lower functioning monkeys had exposure levels "within the range now seen in industrialized nations," whereas the higher-performing control monkeys demonstrated "body burdens below the human average" (269).

Human studies exhibit varied results. For example, Steinbraber points out that while several U.S. studies have "documented few problems" from contaminants, a series of studies conducted in the Netherlands "tell a different story" (271). That research found that breast and bottle-fed babies exposed prenatally to PCBs had many deficits, with higher cord-blood levels linked to "poorer" neurological conditions, "lower psychomotor scores," lower "cognitive abilities," "slower reaction times," and "more signs of hyperactive behaviors and attention problems" (271). When we look at lactational exposures, the picture is more complicated, since breastmilk itself cancels out many of the problems with environmental exposures. Indeed, such effects "were more subtle" than the prenatal effects; at eighteen months, "breastfed babies, in spite of their higher exposure level, scored higher than formula-fed babies on neurological tests" (271). In addition, three-and-a half year olds breastfed as infants performed better on verbal recognition tests than formula-fed children. Babies whose mothers had the highest breastmilk levels of PCBs and dioxins showed compromised scores on movement and muscular activity, with rates comparable to those fed formula. These children also showed poorer attention spans, thus suggesting that both prenatal and lactational exposures seem to diminish a child's ability to pay attention. When these same researchers considered immunological developments on the same children, the effects were more pronounced, with the "breast milk of some Dutch women contain[ing] levels of PCBs high enough to compromise its immune-boosting powers."

It becomes clear to anyone following the issue that discussions—whether they appear in the popular press or scientific journals—almost always follow up a mention of toxicity with claims about breastmilk's continuing advantages. Holding out her own choice to nurse her daughter for two years, Steingraber notes that "if forced to agree or disagree" she would err on the side of breastfeeding (274)). Like Law and the other feminists who downplay science, Steingraber does not embrace all scientific studies and their conclusions without weighing them carefully. For example, she notes that she "believe[s many] risk/benefit analyses are an unhelpful approach to the problem of chemical contaminants in breast milk." Why? Because "they offer no solutions." But while she points to problems with some studies and with risk/benefit analyses in particular, she does not dismiss them outright. As I have suggested elsewhere in this book, risk/benefit analyses are often clumsy; they frequently consider narrow frames of reference, isolating findings from their contexts. As policy decisions they often involve questions about issues other than health and well-being, such as economic considerations. Steingraber notes that an early risk assessment compared "lives saved from infectious diseases with an estimate of the number of additional cases of cancer that might be caused by the exposure to carcinogenic chemicals in breast milk" (274). Concluding that fewer children die from breastmilk induced-cancers than from formula-induced infections, and that therefore "breast is best," this study failed to look at other health effects, including immune functioning, hormone disruption, and altered brain development, all of which are significant given that we are talking about processes that are occuring during periods of acute growth, when organs and tissues are undergoing significant development.

FEMINIST ATTENTION TO THE ENVIRONMENTAL CONTAMINATION OF BREASTMILK

If we return to the question of attention to breastmilk contamination from feminists, we have to acknowledge that answering it involves revisiting what we mean when we use the term "feminist." We might break feminist—here, in its widest sense—attention to the environmental contamination of breastmilk into five categories: first are those who self-identify as "feminists," and who, if they broach the issue, tend only to mention it or address it only briefly (Carter; Hausman; Van Esterick) second are those I will call "Rachel's daughters," who tend to be scientists writing for laypersons, interested in bringing attention to the ramifications of environmental degradation and thus broaching systemic issues leading to environmental problems—but who are concerned foremost with contamination itself (Steingraber; Colborn, Dumanoski, and Myers, the authors of *Our Stolen Future*; and Schettler, Solomon, Valenti and Huddle, the authors of *Generations at Risk: Reproductive Health and the Environment*; third are "breastfeeding advocates" who, like feminists, if they address toxics issues, tend only to mention them or dismiss them (Palmer; Baumslag and Michels); fourth are "ecofeminists" (Mies and Shiva), who consider systemic effects of patriarchy and colonizing approaches to the environment, and if they mention breastmilk contamination, do so in the course of these larger agendas; and fifth, environmental justice activists (Cook; LaDuke) who sometimes place breastmilk contamination at the forefront of their attention, but who oftentimes do not identify as feminist. While such categories are clearly clumsy and problematic, and not exhaustive, what brings these thinkers together is that they possess an understanding of structural issues that we might loosely characterize as feminist, and this is important in our assessment of attention to the environmental contamination of breastmilk.

Clearly, Steingraber offers us much on this issue, as previously described. In *Our Stolen Future*, Colborn, Dumanoski, and Myers—a scientist/writer team—address breastmilk contamination in the course of considering a host of toxics issues. They point out that "researchers are only beginning to appreciate the myriad benefits of breast-feeding, which . . . aids in mother–baby bonding . . . [and] also provides infants with important immune protection and a host of substances that enhance development" (215). Noting the positives, they are more wary than Steingraber, stating that "we know too little to judge how the undeniable benefits of breast-feeding balance against the risks of transferring hormonally active contaminants" (215). Cautioning "There is a pressing need for research to determine whether the concentrations of hormone-disrupting chemicals in human milk pose enough of a hazard to make breast-feeding inadvisable for some women," they conclude that "we cannot afford to ignore the pressing issue of persistent contaminants when weighing the merits of breast-feeding against alternatives such as bottle-feeding with a formula based on cow's milk" (216). Their warnings that people not trust authorities or labels speaks to the kind of approach they take, which has a feminist orientation. For example, they insist: "Never assume a pesticide is safe. Anything designed to disrupt living organisms—plant or animal—may also prove harmful to humans or other animals in unexpected ways" (216).

Asking us to "recall EPA researcher Earl Gray's discovery that products designed to kill fungus on fruits and vegetables can interfere with the synthesis of steroid hormones in animals and most likely in humans as well," Colborn, Dumanoski, and Myers warn that "studies have found higher rates of cancer in children and dogs living in homes that use pesticides in the home and garden," and assert that "the casual use of pesticides around homes and gardens for frivolous, cosmetic purposes is risky and irresponsible" (216). Indicting trusting suburbanites who think that only farmworkers directly involved in applying pesticides have reason to fear, they urge: "Don't be blasé about the risks that come along with pest control"; indeed, "In the United States, greater quantities of pesticides are applied per acre in the suburbs than on agricultural land" (216–17). They remind us that the problems go far beyond pesticide issues, while solutions also are beyond what an individual can do. Colborn's suggestion that, for the most exposed women, "we just don't know" whether the benefits outweigh the risks; her claim that "so far, the benefits seem to outweigh the risks, but we just don't know" brought her into conflict with some breastfeeding advocates, a controversy that played out on the pages of *Mother Jones*.[16] Her calls for "systematic assessments," "changes to laws and regulations," including broadening the Toxic Release Inventory and shifting the "burden of proof to chemical manufacturers," position her as a strong advocate for changes that would involve systemic restructuring of various aspects of our assumptions about governance (219–23).

Ecofeminists Maria Mies and Vandana Shiva mention breastmilk twice in their well-known book, *Ecofeminism*. They use breastfeeding as an example of a "subsistence way" in which women can resist the tendency to commodify and be commodifiers, noting that "such a change of life-style on a large scale would also change the consumption model which the North's middle classes provide for their own country's lower classes and for the people of the South"(Mies "Liberating" 256). Addressing the fact that many German women are alarmed over "the poisoning of mother's milk . . . by DDT and other toxic chemicals," they recount the story of a German woman who wants Germany to import ragi, a nutritious millet that could be used for baby food, from South India, so as to "solve the problem of desperate mothers whose breast milk is poisoned and give the poor in South India a new source of money income" ("Myth" 67). Mies's response, that if "the subsistence food of the poor entered the world market and became an export commodity it would no longer be available to the poor" and would soon be grown with pesticides if the project were to work, becomes, for her, an example of how erroneous he belief is the belief that the workings of the market can solve problems ("Myth" 68). She further argues how such thinking "leads to antagonistic interests even of mothers, who want to give their infants unpolluted food" (68).

This example highlights the complexities of such issues as well as how we must bear in mind the ramifications of decisions about infant feeding on women and on children in diverse geographies. As Vandana Shiva points out, "The child has been excluded from concern, and cultures which were child-centered have been destroyed and marginalized" ("Impoverishment" 88). For her, the world's policymakers face a challenge "to learn from mothers, from tribals and other communities, how to focus decisions on the well-being of children" (88). Shiva concludes "Putting women and children first needs above all a reversal of the logic which has treated women as subordinate because they create life, and men as superior because they destroy it" (88). To foreground the environmental contamination of breastmilk is to confront a series of problems central to contemporary culture—including sexism, racism, classism, ageism, and imperialism—problems that lie at the heart of feminisms' concern with the interlocking nature of various forms of oppression.

FEMINISTS, ENVIRONMENTAL JUSTICE ACTIVISTS,
AND BREASTMILK TOXICITY

Having considered various "feminist" perspectives, I want to turn to the question of the attention environmental justice activists bring to the issue of environmentally contaminated breastmilk. Many who make up its ranks, as well as its leadership, are women coming to activism out of concern for the health of their families and their children. One key example of such activism is Katsi Cook, a Mohawk midwife who, after finding that the animals and plants in her region suffered from high toxicity levels, became concerned about the breastmilk of mothers there and began to draw together women with similar worries. I will discuss the work of these women and others in chapter four, but here I want to look at connections or disconnections between environmental justice activists and U.S.-based feminists about contaminated breastmilk. It is important to note that, at least since the 1970s, women in the United States have been working at the grassroots level to focus attention on the social, political, economic, and ideological events that have brought us to an impasse in which toxics inhabit our air, our soil, our water, our plants, our animals—and our breastmilk. When we think of grassroots organizing around environmental problems, many of us remember Lois Gibbs and her work on the Love Canal incident. Women of color, as well as women of lower economic means, women from certain regions such as "cancer alley" in Louisianna, the California farm belt, or inner-city Los Angeles, as well as women in certain work environments with greater occupational exposures, have been at the forefront of raising public awareness and struggling against corporate and governmental policies that place women and their families at risk.

But while what has come to be called the environmental justice movement boasts of women in leadership and membership positions (Di Chiro; Kirk; Pulido; Seager), scholars have pointed out that race, class, ethnicity, or other marks of positionality, rather than gender, serve to define the group— conceptualizations many of these women bring to their activism (Pulido 19). Some scholars have also noted that the work of such women is not typically informed by a femininist analysis (Pulido 19). Pulido points out that despite living "highly gendered lives," many of these women "often choose *not* to emphasize their female identities, but rather define themselves in terms of 'race' and 'place'" (19). But what does it mean to claim that such women lead "highly gendered lives"? If they tend not to emphasize gender by focusing on the "patriarchy" or "sexist" oppression as something against which they must struggle, it becomes apparent that many who work on behalf of the movement approach their activism from their positions as mothers (Di Chiro; Hamilton; Kirk; Pardo; Pulido).

The environmental justice movement is marked by its expansion of the term "environment" to include not just "green" spaces, so-called wilderness areas (itself a colonizing term), and animal habitat but also "the places where people live." Its struggles involve bringing attention to how we have allowed some people's living spaces to become more polluted and more dangerous than other people's. Toxics struggles are at the center of this movement. And, quite often, these activists approach pollution issues from their positions as mothers. As community activist Cynthia Cockburn notes, "In a housing situation that is a health hazard, the woman is more likely to act than the man because she lives there all day and because she is impelled by fear for her children" (Hamilton 208–9). Cindie Spencer points out that "women throughout the world are often the first to notice and respond to changes in the quality of life in their neighborhoods and to draw connections between declining health conditions and the newest hazardous facility in their community" (4). Once aware of threats to the health of their families, women challenge traditional gender expectations and empower themselves by becoming spokespersons and activists for justice.

Another point that must be made about breastmilk contamination is that women of color and poorer women face environmental toxicity at much higher levels so it makes sense that these would be the women who have brought more attention to the environmental contamination of breastmilk than other environmental justice activists. Breastfeeding rates among other women of color have begun to rise. As this trend continues, we would expect to see breastmilk contamination move up on the environmental agenda.

In their study, "Indigenous Women Activists and Political Participation," Prindeville and Bretting point out that "it is well-documented in the literature that many women of color reject the label 'feminist' " since many perceive it to "privilege the perspective of middle-class Euro-American women" (48). They note that "since both Latinas and Native American women have historically experienced colonization, racism, economic marginalization, and political repression in addition to sexism," they conceptualize feminism differently than mainstream feminists. Their study of sixteen "indigenous" women activists identifies only five as "feminists" (48).[17] Of those who rejected the label "feminist," all articulated ideals that correspond with "liberal feminist" thinking or with other forms of "feminisms"; that is, all expressed belief in equal rights for all, in empowering women, in "the need for changing patriarchal social structures, and in the need to change institutional, structural, and attitudinal barriers" (48–49).

If we look at what feminists who have addressed environmental issues have said, we find that they tend to consider structural injustice and ways in which a gendered analysis helps forge a healthy response. For example, Carolyn Merchant critiques the subordination of reproduction by production under patriarchal, industrialist capitalism. She comments: "The body, home, and community are sites of women's local experience and local contestation. Women experience the results of toxic dumping on their own bodies (sites of reproduction of the species), in their own homes (sites of the reproduction of daily life), and in their communities and schools (sites for social reproduction)" (*Earthcare* 161). Prindeville and Bretting find that the indigenous women activists whom they interviewed support a "communitarian form of feminism that reflects the culture and belief system of indigenous peoples"; in other words, "Their support of equality and access to opportunities encompasses all members of the community, not just women" (50). In addition, while studies have shown that Native American activists tend to see the environmental pollution of their communities as systematic genocide, other women of color and working-class women conceptualize a distrust that may or may not involve viewing toxics issues in terms of "genocide." What brings such activists together is a distrust of those in decision-making positions along with a vision that systemic restructuring must occur for justice to prevail (Hamilton; Krauss; Prindeville and Bretting). These activists share much with "mainstream" feminists, so that an opening for coalition work clearly exists. As Frendenberg and Steinsapir point out, "By raising health concerns and by linking environmental issues to struggles for social justice and equity, the grassroots environmental movement has created the potential for a cross-class movement with a broader agenda, more diverse constituencies and a more radical critique of contemporary society than that of the national environmental organizations" (242). Clearly, some believe the same can be said of "feminist" organizations. Others envision coalition work that dissolves explicit group identity.

Peggy Antrobus, a founder of Development Alternatives with Women for a New Era (DAWN) articulates these common bonds. "We are different women, but women nonetheless. The analysis and the perspectives that we get from women are certainly mediated by, influenced very profoundly by differences of class, and race, and age, and culture, and physical endowment, and geographic location" (quoted in Seager *Earth* 280). But, she notes, the commonality that women share, including the commonality "in our vulnerability to violence . . . in our otherness, in our alienation and exclusion from decision-making at all levels" can be a catalyst (280). Feminist poet and commentator Audrie Lorde's reminder that it is not just commonality, but difference, that offers the strength needed to bring about creative change is a way of thinking about how women (and their allies) of vastly differing experiences can come together. Noting that "[t]he old definitions have not

served us, nor the earth that supports us," that "our future survival is predi-
cated upon our ability to relate within equality," she emphasizes the power
of working with our differences, embracing them ("Age" 123). In her always
powerful, always beautiful voice, she reminds us: "Now we must recognize
differences among women who are our equals, neither inferior nor superior,
and devise ways to use each others' difference to enrich our visions and our
joint struggles" (122–23). Why? Because "[t]he future of our earth may depend upon the ability of all women to identify and develop new definitions
of power and new patterns of relating across difference" (123).

STRUCTURAL PROBLEMS IMPEDING BREASTFEEDING IN THE UNITED STATES: WOMEN IN THE WORKPLACE AND IN ENVIRONMENTALLY DEVASTATED COMMUNITIES

If we concede that using breastmilk confers palpable benefits to both mother
and child, as well as to societies and the larger ecosystem, and that the longer
one nurses, the more the benefits, the next step would be to assess what forces
impede various women's choices about infant feeding. It is instructive to consider some of the history of policies that impact women's work outside the
home since this has some bearing on policies governing women's ability to use
breastmilk after they have returned to work. Not surprisingly, fears about calling attention to women's "difference" from the "traditional" male worker have
surfaced in debates around the construction of legislation regarding women's
employment. This has increasingly become an issue in the United States, as
more and more mothers with young children have entered the workforce. As
Judith Galtry suggests, tracing various reactions by the National Organization
for Women (NOW)—serving as a representative of liberal feminist priorities—to developments concerning mothers in workplace geographies highlights the tensions that characterize how this issue has played out in the
context of U.S. social, political, and legislative histories ("Extending" 297–98).
As Galtry points out, attempts "to minimize the costs of gender difference
associated with women's traditional childbearing/ rearing roles" have meant
that "debates [have] centered on the relative merits of an equal-treatment and/
or special-treatment approach to the incorporation of women's reproductive
requirement within an emerging equality and workplace requirement" ("Extending" 296). Since its birth in 1966, NOW's dedication to the liberal values
of equality, liberty, and justice, as well as the importance of gender-neutral
legislation promoting equality in paid employment and politics, has meant that
NOW has rejected gender-specific protective labor legislation in favor of equal
treatment approaches (297; Vogel "Debating" 78).

    And so, the 1978 Pregnancy and Discrimination Act (PDA) requires
that pregnant women be treated in the same manner as other employees
with similar abilities or limitations, while the 1993 Family and Medical
Leave Act (FMLA) provides twelve weeks of leave for a range of medical

and family reasons not distinguishing between reasons women and men might need to take time away from work (Galtry "Extending"; Kamerman, Kahn, and Kingston). Breastfeeding, therefore, has not tended to figure into these debates or has been systematically downplayed or neglected (Galtry "Extending"; Taub; Taub and Williams). This helps explain the current dearth of feminist advocacy around breastfeeding issues, including breastmilk contamination. Invoking the figure of breastfeeding clearly represents a risk to many feminists. Hill Kay's argument that we need to draw "a bright line" between the "female specific functions of pregnancy and childbirth, and the potentially gender-neutral practice of child rearing" (in which she includes breastfeeding) suggests that lack of feminist attention to breastfeeding stems, in part, from the assumption that breastfeeding is not necessary and too costly in terms of equal treatment discourses (quoted in Galtry "Extending" 299–300). That Jules Law makes a similar argument about the "seemingly inevitable slippage between reproductive and social-reproductive issues, between child bearing and child rearing" corroborates this stance (407).

But the release, in 1997, of a policy statement by the American Academy of Pediatrics (AAP) strongly recommending that all infants be breastfed for twelve months and calling on employers to support breastfeeding meant that careful focus on establishing gender-neutral policies would become challenged.[18] Significantly, while advocacy groups such as La Leche League had made many of the pro-breastfeeding arguments used by the AAP for years, the AAP's statement plays a significant role in bringing the attention of policymakers and groups such as NOW to the table, thus marking an important event in tracing the trajectory of how infant feeding gets framed. Soon after the recommendation's release, New York Representative Carolyn Maloney's announcement of her intent to introduce legislation supporting breastfeeding for paid workers resulted in the New Mothers' Breastfeeding Promotion and Protection Bill (H.R. 3531) in March 1998 (Congresswoman). NOW's initial reaction, in which a spokesperson for a local chapter dismissed the AAP's recommendation on the grounds that it "places pressure on women to breastfeed" and thus is "anti-woman," hints at the tensions involved, where dictates that women should breastfeed are weighed against what has amounted to "dictates" in the workplace, that women cannot simultaneously use breastmilk and continue working (Galtry "Extending" 297). It also suggests the assumption on the part of NOW that whether or not breastfeeding limits women's progress as "equal" participants in the workforce, it is perceived to do so and thus is dangerous.

As Galtry explains, after that initial period, NOW reversed its position and lodged a press release supporting nursing (also see Gartner 19–20). NOW's subsequent reaction, an official statement called "NOW Appreciates AAP's Recommendations on Breast Feeding, Calls on Business and Society to Support Findings," also reflects ambivalence and tensions, although less so than its initial response ("NOW Appreciates"). Stating that "when women do not have to hide in the bathroom or in a corner to breastfeed or pump, we will have come a long way toward real respect for the job of being a mother," it also urges that we "put the AAP's recommendations in perspective," and regard infant feeding as a "personal decision" with differing factors for differing individuals ("NOW Appreciates"). Wary of wholehearted support for the recommendation, these documents argue that "some women find it very difficult to breast feed because of financial, logistical, health, or other reasons" and that "we should not use these findings to judge some mothers as 'good' and others 'bad' because of their decision on this one issue of their baby's health care." The minimizing of infant feeding as one issue among many healthcare issues, the lack of attention to the social situatedness of perceived "difficulty," and the emphasis on choice, which, according to Galtry, "caused a furor" among breastfeeding advocates, mirrors similar emphases on choice in earlier editions of the Boston Women Health Collective's *Our Bodies, Ourselves*. For example, the 1978 edition claims that "the propagandists would have us believe that if we breastfeed we are good and if we bottle feed we are bad."

Like the initial NOW dismissal of the AAP recommendation that all women breastfeed for a year, a dismissal later pasted over by approaches suggesting that breastfeeding may not be all about coercing women to risk hard-won moves toward equality in the workplace, the early *Our Bodies, Ourselves* seems hesitant about letting go of the construction of breastfeeding as one issue among many.[19] Like many contemporary feminist commentaries, they resist pro-breastfeeding strategies because they view them as coercive and burdening women with restrictive demands. Subsequent NOW statements (e.g., its press release "NOW Demands Greater Acceptance of and Access for Breastfeeding Mothers") take a less equivocal approach ("Now Demands"). These support the bill and lay out NOW's recommendations to policymakers, including requests that governments provide programs for low-income mothers to have access to breast pumps and extended nutritional supplements through WIC (women, infants, children) programs, and that they increase funding for workplace childcare, and issue federal guidelines outlining practices to accommodate breastfeeding mothers ("NOW Demands"). Similarly, later editions of *Our Bodies, Ourselves* reflect a change from emphasis on choice toward greater problematizing of choice constructions. The 1998 "newly revised and updated" edition, for example, notes in its list of "myths" about breastfeeding that "we may hear that nursing . . . ties us down so that we can't resume work or other activities," suggesting that this is not the case (507).

Feminists, environmentalists, and environmental justice activists have increasingly been bringing attention to environmental injustice and the ways in which toxicity issues mirror larger structural issues both within the United States and globally. That all these groups have also been moving toward coalition work as well as implementing new strategies such as participatory action research suggests that we have reason to hope that this issue will continue to garner the attention it deserves.

FOUR

POLLUTING THE "WATERS"
OF THE MOST VULNERABLE

ENVIRONMENTAL RACISM, ENVIRONMENTAL JUSTICE, AND BREASTMILK CONTAMINATION

TOXIC BREASTMILK AS AN "ENVIRONMENTAL JUSTICE" ISSUE

Due to the relentless work of countless environmental justice activists, more and more people are beginning to understand what we mean when we talk about environmental racism and environmental in-justice. And many are beginning to realize that these labels refer to something very real going on in our midst. Furthermore, when we consider what environmental in-justice involves—the inequitable distribution of the costs of current practices—and when we think about the realities of breastmilk toxicity, we see that the environmental contamination of breastmilk is an environmental justice issue. Why? And why is it useful to make this claim? The short answer involves the recognition that certain groups of people—those less privileged because of their "race," their ethnicity, their "color," their gender, their class status, or other markers of social stratification—get situated in such a way that they find themselves more succeptible to environmental pollutants in general. The environmental contamination of breastmilk follows many of the same patterns that characterize more general issues of environmental injustice.

But the contours of the story surrounding the environmental contamination of breastmilk and how it becomes intertwined with other environmental justice issues takes some circuitous paths, involving such questions as "what groups bear high rates of environmental contaminants in their breastmilk?," "what groups have been working on this issue?" "what kinds of attention have they brought to it?" "have they framed it as an environmental justice issue?, a gender issue?, an issue of rights?" "what kinds of silences surround the environmental contamination of breastmilk?" and "what kinds of discussion/work still need to take place?" Many who follow in Rachel Carson's steps, working to end further devastation of our environment, adopt the image of the canary in the mine to suggest the ways in which we can be likened to the brightly colored, twittering creature. The image reminds us how plant and animal sentinels function as powerful warning signals, indicating when human beings might be at risk, but also illustrates that we are like that canary, and that the most vulnerable among us occupy this position first. Indeed, the image marks for us the ways in which some peoples (like some species) bear the effects of environmental toxicity more pointedly than others, while also hinting at the degree to which all species are interlinked with each other and with various ecosystems. In the words of Daniel Quinn, when one rivet gets pulled out, things might seem OK, but when several are pulled, it may be too late for us, attempting to fly our figurative airplane, to return to safety (*Ishmael*).

Paul Hawken, quoting a piece by two scientists, Weil and Kirchner, in the 2000 issue of *Nature*, looks at extinction and biodiversity, reminding us of the ways in which we are all connected, and of the fact that loss that comes about through environmental degradation has huge ramifications (1). The *Nature* article notes that "when you lose a species it's not ever coming back. . . . extinction is final." Musing on "mass extinctions," and how "it takes the earth ten million years to recover" from one of these, it goes on to state that "it takes the environment just as long to recover from the extinction of even a few species, small events which nevertheless rip holes in the biosphere that are impossible ever fully to repair" (quoted in Hawken 1). Then comes the quotation that becomes even more poignant when we think of the loss for our children when we pollute the earth to such a degree that they are denied so many things—including the right to be nursed and to choose to nurse their own children. It reads: "If we deplete the planet's biological diversity, we will . . . leave a biologically impoverished planet, not only for our children or our children's children, but for all children of our species that there ever will be" (1). The finality of that reminder, the thought of the cold, stark truth of what could become lost if we continue on our current path, rings clear when we begin to add up the list of things our children will be asked to forgo.

Breastmilk is like the canary in the mine. Scientists often use samples of breastmilk to determine the condition of a local ecosystem because breastmilk contains pollutants that are not present in blood and urine samples. Toxic breastmilk thus becomes, not just a tragedy in itself, but an indication of much more wide-ranging devastation. Those conducting research into breastmilk toxicity, however, still suggest that except in the most extreme cases women should nurse their children; the positive effects of breastmilk continue to tip the balance. But what is clear is that unless we come together—locally, as a nation, and as a global collective—and begin addressing the complex set of issues that add up to practices that pollute the environment, the scales will tip the other way. Women will no longer have the choice to nurse their babies, if they want to protect them from the toxins that will have accumulated in the fatty tissues of their bodies. And what is also clear is that, at this point, some women already face such a dilemma. Some women, research has shown, if they want to protect their babies from these dangerous pollutants, must either forsake the choice to nurse, or must express milk and dispose of it, until tests show that their bodies have been cleansed of enough of the chemicals to make nursing advantageous. And, by and large, the women who have to make these choices are Native American women, women of color, and poor women. The case of breastmilk toxicity spells out loud and clear that sexism, racism, and classism continue to haunt the lives of people in the United States and the world over.

Felix Cohen, "the founder of federal Indian law," adopts the image of the canary in the mine in discussing the effects of environmental toxins on Native Americans, widely recognized as among those most subjected to environmental hazards (Weaver 2). He points out that "like the miner's canary, the Indian marks the shift from fresh air to poison gas in our political atmosphere; and our treatment of Indians, even more than our treatment of other minorities, marks the rise and fall of our democratic faith" (quoted in Weaver 2). Observing that "Native America today provides a virtual catalogue of environmental destruction" Donald A. Grinde, Jr. and Bruce Johansen suggest the degree to which Native American peoples continue to bear the brunt of environmental destructiveness; they and other vulnerable groups become harbingers of what the rest of us—even the most privileged—can expect to encounter if we continue on our current path of thoughtlessness (3). The canary image reminds us that breastfeeding infants, like various species of fish, birds, and other wildlife, as well as Native American peoples, occupy a position of vulnerability. Buluga whales, salmon, and other species often exhibit the effects of environmental pollutants before others because they inhabit the dark, watery spaces, the rivers and the seas where we allow the waste from farming and polluting industries to seep and cause mutations in the delicate and vital systems of creatures that form part of the vast foodchain.

Breastmilk is like that. Like the rivers and the oceans, it becomes a repository for chemicals that cannot be broken down by the body. And our infants and fetuses, first through contact with the watery placental fluid, and then through the vital fluid imparted by the mother, become subjected to the pollutants that we allow to collect in the watery spaces of our communities and our bodies.[1] Rachel Carson warned so long ago: "It is not possible to add pesticides to water anywhere without threatening the purity of water everywhwere" (42). Beckoning us to take heed, she explains, "Except for what enters streams directly as rain or surface runoff, all the running water

of the earth's surface was at one time groundwater . . . [so that] in a very real
and frightening sense, pollution of the groundwater is pollution everywhere"
(42). And just as all water flows into all other water, with pollution infecting
the whole system, when a woman's breastmilk gets contaminated with certain
persistent organic chemicals, those chemicals remain not only in her body, but
are passed to her daughter, and persist in her daughter's body until her daughter breastfeeds her own infant. It is important that we remind ourselves that
human infants, like other vulnerable populations and species, are particularly
susceptible to environmental contamination. During significant periods of
development—when the child is still in the womb or in the early stages of
growth, what doctors call critical windows—its systems undergo periods of
rapid growth; these are the moments when environmental contaminants can
do the most harm (Center; Landrigan; National Academy of Sciences).[2]

Another connection between the toxicity we have come to associate
with the environmental justice movement and breastmilk contamination
concerns our approach to the dark, watery spaces—the oceans, rivers, lakes,
and seas. These are the spaces where environmental contaminants seep and
get released. And like the milk of mothers, these spaces, because they seem
dark and other, because they are out of sight, get forgotten and ignored. And
it is no accident, this connection between the dark, watery spaces, the milk
of mothers, and the milk of certain mothers, the darker ones, the ones who
are forgotten, those living closer to landfills and polluting industries, those
relying on fish for subsistence, those living in barrios, in unsafe inner city
environments, or on reservations. This chapter considers some cases of environmental injustice, the effects on breastmilk, and some of the work that
has been done to address this inequity.

THE ENVIRONMENTAL JUSTICE MOVEMENT: A RANGE OF
DOCUMENTED CASES, WITH THE LIST STILL GROWING

Researchers, scholars, activists, and policymakers have increasingly focused their
attention on the environmental justice (EJ) movement, which has grown up
around the reality that communities of color, Native American peoples, and
those of lower economic standing confront "environmental stressors" at a rate
much higher than those who make up the dominant culture (Lee 141). While
researchers continue to document the negative health impacts of the "disproportionate exposure to environmental and occupational toxins," which are exacerbated by stressors in the physical and social environments, the "emerging
literature . . . has begun to conclusively document serious environmental inequities" that give rise to these health impacts (Lee 141). These include "lead poisoning; air pollution and ambient air quality; groundwater contamination and
drinking water safety; proximity to noxious facilities, mining waste and nuclear
plants; location of municipal landfills, incinerators, and abandoned toxic waste
sites; placement of transportation thoroughfares; illegal dumping; occupational

health and safety; use of agricultural chemicals; contaminated fish consumption; habitat destruction; cleanup of Superfund sites; and unequal enforcement of environmental laws" (Lavelle; Lee 141; Sexton; U.S. Department of Agriculture; U.S. EPA "Environmental Equity"; Wasserstrom; Wernette; West).

As leading EJ activist Robert Bullard tells it, "The environmental justice movement took shape out of the 1982 protests in Warren County, North Carolina" ("Environmental" 5). In that incident, officials selected the mostly African American county as the site on which to dump 30,000 cubic yards of soil highly contaminated with PCBs. As in countless similar situations, opponents argued that "the decision made more political sense than environmental sense"; the site choice had to do with an official position of placing less value on the health and lives of these citizens than on others considered more "important" (5). These protests, in which "regular folks" turned activists lay down on the highway to block the dump trucks, a scene depicted on the front cover of Bullard's book *Dumping in Dixie*, mark a significant episode in the environmental justice movement as one of the first times "African Americans had mobilized a national, broad-based group to oppose what they defined as environmental racism" ("Environmental" 5). And 1987, when the United Church of Christ's Commission on Racial Justice completed its study and report titled "Toxic Wastes and Race in the United States," is now heralded as the year in which we started to talk about the environmental justice movement as a phenomenon.

The next big event in the movement's history occurred in 1991 when many people and groups came together in Washington, DC for the First National People of Color Environmental Justice Leadership Summit. Also, in that year, a group of EJ advocates led by Benjamin Chavis contacted the "Big Ten" environmental groups, arguing that they had been racist in their hiring of staff and their priority setting (Alston 29). Until then, most people associated the environmental movement with "green" issues such as wildlife preservation, conservation, global warming, species depletion, or preserving the rain forests. With the growth of the EJ movement, more people began to think of the term "environment" in an expanded light, realizing that the "environment" is where we live, including urban environments. Under this expanded definition, issues such as occupational safety, toxic waste disposal, trash incineration, the management of industrial and mining activities, pesticide production and use all could also and should be considered "environmental" issues. People began to understand that one's race, ethnicity, and class status have a lot to do with the degree to which one will be subjected to environmental pollution. Indeed, the UCC Commission found that "[r]ace proved to be the most significant among variables tested" and that "this represents a consistent national pattern" (Grossman 277). It also recognized that "although socio-economic status appeared to play an important role in the location of commercial hazardous waste facilities, race still proved to be more significant" (277). Furthermore, the commission documented that, at the time of its study, "three out of the five largest commercial hazardous waste landfills in the United States were located in predominantly black or Hispanic communities," while "three out of every five black and Hispanic Americans lived in communities with one or more uncontrolled toxic waste sites" (277). In addition, a 1992 study by staff writers from the *National Law Journal* disclosed inequities in the way the federal EPA enforced its laws; for example, it revealed that penalties under hazardous waste laws at sites with greater white populations were 500 percent higher than penalties with greater populations made up of people of color, that for all federal environmental laws designed to protect citizens from various polluting sources, penalties for white communities were 46 percent higher than those in communities of color, while abandoned hazardous waste sites in those areas take 20 percent longer to be placed on the Superfund national priority list than sites in areas where white residents live (Bullard "Environmental" 9–10).

These figures still haunt our system. And they continue to jolt us into recognition that we are all implicated—even those of us who have long recognized the systemic injustices that oppress those who do not fit the profile of the "norm." It is time for us to recognize that the most vulnerable, particularly the infants and children of those already bearing greater burdens, need our attention, both because they are like canaries in the mine and because it is shameful to forget the canary and the story it tells. Carson's title *Silent Spring* poignantly reminds us that the spring is not only the season for songbirds to hatch and begin their chirping, but that it is also the symbolic time of renewal, the time of birth, when young birds begin to hatch from their fragile shells and when babies begin to exit the womb.

CHEMICAL COLONIZATION: TOXICITY ON THE AKWASASNE

Perhaps the most telling case of infant food toxicity involves Native American peoples living on reservations where environmental waste contaminates the water, the air, the soil, the plants, the fish and wildlife—and the breastmilk of mothers. Repeatedly and systematically, Native American communities have been targeted as sites for locating polluting activities (Grinde and Johansen; La Duke; Moraga; Weaver). Native American activist and Ralph Nader presidential running mate Winona LaDuke points out that a Worldwatch Institute study found that 317 reservations in the United States "are threatened by environmental hazards, ranging from toxic wastes to clearcuts" (*All* 2). Indeed, devastation "can be seen in nearly every quarter of Indian country—in the Navajo Four Corners region, the coal strip-mining operations on Cheyenne and Crow Lands in Montana, Pyramid Lake in Nevada, the Arkansas River-bottom ripoff in Eastern Oaklahoma" (Hobson 3). As of 1994, seventy-seven sacred sites had been disturbed or desecrated through resource extraction or "development" (3). In addition, La Duke notes, "[o]ver 100 proposals have been floated in recent years to dump toxic waste in Indian communities" with reservations targeted as nuclear waste dumping grounds in sixteen proposals (*All* 2). The choice of the Shoshone people's Yucca Mountain in Nevada as the receptor of high-level radioactive waste shipments of 77,000 tons from more than 100 military and civilian nuclear reactors from around the country constitutes a recent chapter in a long history of oppression and disrespect toward Native American peoples (Mitchell A 14). Andy Smith, arguing that such events amount to "genocide" requiring us to embrace an "anticolonial framework," adds that military and nuclear testing takes place on native lands, and that there have been at least 650 nuclear explosions on Shoshone land in Nevada, with 50 percent of the underground tests leaking radiation ("Ecofeminism" 23; Tallman 17).

While such stories rarely find their way into the mainstream media, La Duke and several Native American activist groups have been speaking out. For instance, she and a few others brought attention to the Mohawk people of the Akwasasne, part of the Haudenosaunee, or Six Nations Iroquois Confederacy, "which is among the most ancient continuously operating governments in the world" (La Duke *All* 12).[3] It is also located on one of the most highly contaminated waterways. Today, around 25 percent of all North American industry is situated on or near the Great Lakes, all of which are drained by the St. Lawrence River, home to the St. Regis Mohawk of the Akwasasne (Thomas 1). Downstream from the lethal sludge of the Great Lakes, Akwasasne also hosts several polluting industries: General Motors, Reynolds, Alcoa, and Domitar (La Duke *All* 15). Dr. Lennart Krook and Dr. George Maylin, two veterinarians, tell of how "Mohawk farmers suffered severe stock losses of their dairy herds in the mid-1970s due to poor reproductive functions and fluorosis, a brittling and breakage of teeth and bones. . . . Additional studies have shown that vegetation suffered as well" (15). This began a quest for measuring the effects of these industries.

Finally, in 1979, PCBs, known to cause liver, brain, nerve, and skin disorders in people, and shrinking testicles, hormone disruption, cancer, immunological, epidemiological, and reproductive disorders in animals—and which saturated the water, soil, and air in the region—were banned (Colburn; Schettler, Seely 2). In 1981, the New York State Department of Environmental Conservation (DEC), reporting "widespread contamination of local groundwater" by PCBs, lead, chromium, mercury, cadmium, and antimony, finally began to target General Motors. In 1983, the parties fined GM under the Toxic Substances Control Act ("Akwasasne Task" quoted in La Duke 17). Using the language of chemical trespass, La Duke complains that such resolve "quickly eroded" (17). For example, in August 1990, the EPA balked, allowing that "containment" as opposed to "treatment" can be acceptable for sites contaminated with PCBs between 10 and 500 parts per million, a change meaning that the company would have to treat only 54,000 of the original 171,000 cubic yards of soil designated "contaminated" (18). Dave Arquette, a specialist with the tribe, notes that "in one core sample of the river bottom at the GM site we tested, we found over 6,000 parts per million (ppm) of PCB's" (18). The tribe's standard is .1 ppm. But instead of finding remediation, they tell of the day when specialists, "covered head to toe in white spaceman-like suits," arrived at the site now called contaminated cove and "covered it with an impermeable sheath," meant as a temporary remedy. Today it still remains without a lining, real containment, or remediation (Sengupta B1).

The story of the men in white suits and the plight of the Mohawk appeared in the "Metro" section of the *New York Times*, a somewhat unusual event in that the media has remained for the most part quiet about the contamination of native lands and infant nourishment. It includes a narrative with commentary by the Thompsons, whose people have lived on the site for centuries. Paul Thompson tells of how his children played and "foraged in" the heap, "pluck[ing] scrap metal" to sell and wood to burn at home (B4). He recounts how "Mohawk from other parts of the reservation rowed down the river to get oil drums, rinsed them and used them to collect rain water"—"recycling," he comments. Ms. Thompson poignantly remembers, "When they put the cap on it, they had their men in white suits. White suits! Our kids are riding on three-wheelers out there and they never even notified us. I ask myself—why? Because we are just native people? I look at that every day and I get so angry." Chris Amato, an assistant attorney general working on the case, agrees, claiming: "I guarantee you if this site was located next to a very middle-class, white neighborhood, this site would be well on its way to being remediated" (Thomas 2). And the famous Love Canal incident, though not a middle-class story, but primarily a working-class one, suggests that this is true. Giving voice to what Audrie Lorde has called "the uses of anger," "anger expressed and translated into action in the service of our vision and our future," anger "loaded with information and energy," the Mohawk women have responded (125).

THE MOTHER'S MILK PROJECT

Their story begins in the mid-1980s when Katsi Cook, a Mohawk midwife, and other Mohawk women determined to find out what was happening in their bodies.[4] At this time, Mary Arquette, who was working as an intern with Ward Stone, a pathologist with the New York Department of Environmental Conservation and a committed voice of warning, was testing a snapping turtle in the region and found that it contained a staggering 2,000 ppm, while the "standard" for edible fish is 2 ppm (Martin 2). Stone, in work documenting toxicity in beluga whales of the St. Lawrence, had found that these whales "carry some of the highest body burdens of toxic chemicals in the world and suffer from a host of problems, including rare cancers and pronounced disease and mortality among young whales" and have a "reproductive success-rate one-third that of belugas in the Arctic Ocean" (La Duke *All* 19; Grinde and Johansen 171–203). This mirrors the health of Native

American youth throughout the United States. Noting that "about 60 percent of the energy resources (i.e., coal, oil, uranium) in this country are on Indian land" as well as two-thirds of uranium production, Andy Smith points to studies demonstrating that "Indian people face skyrocketing incidents of radiation poisoning and birth defects" (23). Navajo children growing up in the Four Corners and the Black Hills, she notes, "are developing ovarian and testicular cancers at fifteen times the national average," while Native women on Pine Ridge face a miscarriage rate six times the national average (23; Flynn 1; Tallman 17).

Two factors are important to the story of the Mohawk women: the traditional positions of the environment, the waterways, and the fish in the lives of the Mohawk people, and the process of bioaccumulation (Cook). Ken Jock, director of the tribe's Environmental Division, explains: "The environment here has a different significance than in mainstream society. . . . The people here feel the environment is their mother and depend on it to be healthy" (Martin 3). When high rates of toxins determined that fishing in the area had to be banned, people felt the effect acutely and in many aspects of their lives. Jim Ransom, executive director of the Haudenosaunee Environmental Task Force, observes that "respect for the natural world, that's the core of the community and that's what's being threatened. . . Fishing, for example, has a host of Mohawk words and traditions attached to it. With fishing all but nonexistent, the use and practice of those words and traditions has disappeared as well" (3). When Mary Arquette asks, "How do you replace the tradition and the language associated with fishing? It's a living language, it has to be used in context," she voices the poignancy of toxic fish and polluted water, soil and wildlife to a people who "have traditionally drawn not just material but spiritual sustenance from the land" (Martin 3; Thomas 1).

In the past, walleye pike, perch, bass and sturgeon formed a large part of the diet of the people. And when parents taught children to fish from the rivers, to hunt and trap in the forests, and to live off the land, cherishing it and honoring the gifts from the water and the soil, they were passing down a rich cultural heritage. They participated actively in a sacred drama enacting relationships between the people and the surrounding world. But when Katsi Cook and the other women began their investigations, they came up against bioaccumulation and biomagnification. Cook and the other women who formed the Mother's Milk Project and the First Environment Project, knew that breastfed babies sit at the top of the food chain. She recalls:

> The fact is, women are the first environment. We accumulate toxic chemicals like PCBs, DDT, mirex, HCBs etcetera, dumped into the waters by various industries. They are stored in our body fat and are excreted primarily through our breastmilk. What that means, is that, through our own breastmilk, our sacred natural link to our babies,

they stand the chance of getting concentrated dosages. We were
flabbergasted. (La Duke *All* 17)

The document makes visible the sexism, the ageism, and the colonizing
implications of breastmilk contamination.

The environmental contamination of Mohawk women's breastmilk
constitutes one more incidence of genocidal and colonizing gestures. Schell
and Tarbell point out that "as both citizens, and wards of the U.S. government," Native Americans "are subject to extraordinary interventions by the
U.S. government in their daily lives," including repeated breaking of treaties,
the "forced 'education' and religious assimilation of children—well into the
twentieth century—at boarding schools far away from home," with the "extraction" of children accomplished by the withholding of food rations from
families (838) and the management of land, extraction, and industry activities in such a way that native peoples bear disproportionate costs of current
practices. Indeed, recounting a history of scientific attention to Native
Americans, Schell and Tarbell point to how nineteenth-century scholars
"brought with them all the intellectual baggage typical of the era, including
racial typology and eurocentric hierarchies for classifying government, art,
religion, technology, medicine, etc." (837). Seeking to situate a study with
a methodology developed to address past injustices, they point out that Native
American peoples have been "subjected by greatly varying extents, to a
eurocentric and often blatantly racist representation of themselves to the
scientific world and indirectly to the public at large" (837).

Cook makes clear the importance of breastfeeding to many Native American women subjected to high levels of contaminants. Noting that "[s]ome
scientists are saying our women shouldn't breastfeed," she disagrees, and points
out that "virtually every woman on the planet carries a body burden of PCBs
and other toxics in her breastmilk. It is a potentially alarming issue coming at
a time when at a community level, we are doing everything we can to get
mothers to breastfeed again. For their health and the health of their children"
(La Duke Interview 16). Cook articulates one of the key implications of the
argument that native women should stop nursing. "If that argument gets carried any further, it says, 'we might as well tell them to stop having babies,' and
from an Indian point of view, that's totally unacceptable." Bearing in mind a
comment by Dr. Allan Cunningham, Associate in Clinical Pediatrics at Columbia University, that he "would expect 77 hospital admissions for illness
during the first four months of life in every 1000 bottle fed infants" as opposed
to five hospital admissions for breastfed babies, the Mother's Milk Project's
literature articulates some of the problems associated with a decline in the
practice of breastfeeding. It notes "Babies raised on substitutes to mother's milk
risk digestive problems, allergies, and have an increased risk for developing
diseases in adulthood such as obesity and diabetes" (16). In addition, it notes
that mothers who do not nurse "risk another pregnancy, and a shortened birth

interval . . . In a community with poor water quality, a very high unemployment rate, and a poor economic status, bottle feeding may be an invitation to
tragedy. It is also a great deal more expensive."[5] Compounding these issues are
problems resulting from the abrupt halt in subsistence activities that had offered sustenance and meaning to the lives of the Mohawk people. Ongoing
studies have demonstrated negative health effects of the curtailment of fish
consumption and other traditional food sources (Schell and Tarbell). In addition, shut out from farming, fishing, and trapping, many Mohawk men have
been forced to leave the reservation for work in construction and other jobs;
family dislocation is a reality. The effects of replacing traditional diets with
affordable, standard foods have been pronounced, affecting much more than
just physical health (Arquette; Arquette, Cole, Cook, "Human Health Studies" et al.).

In an interview with Winona La Duke, Cook speaks of the implications of our environmental policies that leave the most vulnerable exposed.
She articulates the almost unbelievable short-sightedness of allowing such
trends to continue: "there may be potential exposure of our future generations" (La Duke Interview 15). Pointing out that "the analysis of Mohawk
mother's milk shows that our bodies are in fact a part of the landfill," Cook
speaks to the racism, the sexism, the ageism, and the colonizing attitudes
that allow this to happen.

THE CONTINUING LEGACY OF CHEMICAL COLONIZATION: THE CASE OF LATINA FARMWORKERS

In the case of farmworkers and their exposure to pesticides, the situation
surrounding breastmilk contamination is much less clear. Little has been
done to broach the subject, either in terms of the documentation of harm to
farmworker children as a result of exposure through breastmilk or in terms
of work to end such exposure. The people who work on farms and produce
the food and other products consumed by U.S. citizens and our trade partners are primarily ethnic minorities. In 1986, of the approximately two million
hired farmworkers in the United States, 90 percent were "people of color"
(Perfecto 180). Although these workers come from a range of locations and
racial and ethnic groupings, including black North Americans, black
Caribbeans, Puerto Ricans, Laotians, Filipinos, Koreans, and Vietnamese,
Latinos of Mexican origin constitute the largest group (Martin et al.; U.S.
Department of Agriculture). In the mid-1980s, they made up 75 percent of
the agricultural workforce, whereas today 95 percent of those workers are
estimated to be Latinos of Mexican descent or those born in Mexico (Acury
235; Perdito 180; "Report on Mt. Olive" 2). Not many have recognized the
gender component of the farmworker story. And, as political scientist Jill
Gay points out, "few studies of pesticide exposure have been done concerning women" (2). Because farm women often get classified as "farmworker's

wives" or "farmer's wives," rather than as "farmworkers" or "farmers," those who frame studies often exclude them from analysis (Gay 2). Women's contact with pesticides through occupational exposures as well as through prescribed gender roles exacerbates this situation. Women face exposures when they wash the clothing of their husbands, partners, sons, and daughters, and when they reside in homes that are often adjacent to fields and subject to aerial spraying (Gay). In addition, women have increasingly begun to work in the fields, in direct contact with pesticides. Indeed, 1994 estimates for the United States suggest that women then made up one-fourth of all farmworker forces (Dwyer; Gay; Ortiz). Two scientists from the Centers for Disease Control (CDC) note that "[m]ost information on the health effects of exposure to hazardous substances comes from occupational studies of healthy males" (Burg 160). Because exposures to women have received little attention, and because not much has been done to recognize the health effects of such exposures, the story of the effects of breastmilk contamination among farmworker women continues to be untold.

Another story that broke recently, about Wal-Mart keeping its prices so low by hiring contracted janitorial workers (many of them Latinos from Central America) who were paid pitiful wages, denied benefits, and made to work, in many cases, seven days a week with no days off, sometimes for years, reminds us of the degree to which things that we often take for granted come at a cost (Lohr Sect 4.1). Crop production in the United States follows similar patterns. As many have noted, while the United States prides itself on offering the cheapest food supply to its consumers, this, too, comes with a cost ("Report on Mt. Olive" 1).

The farmworkers who sow the seeds, work the fields, bring in the harvests, and apply the pesticides bear this cost through their poverty, their misery, and their faltering health. But unlike the case of the illegal exploitation of the Wal-Mart janitorial staff, the harsh conditions endured by farmworkers are legally sanctioned ("Report" 1). In 1960, CBS released a documentary, "Harvest of Shame," that exposed the conditions faced by migrant farmworkers and their families, and brought about some improvements in the working conditions of these laborers ("Report"). Many point out, however, that these gains, which many attribute to the United Farm Workers of America, a California-based labor union, have slowed or been lost, so that now farmworkers continue to confront a host of problems that we do not see when we make our trips to the grocery store. Indeed, "Agricultural economists and industry surveys found [in 1997] that wages ha[d] fallen by as much as twenty percent" in the two decades preceding those surveys, and continue to fall today ("Report" 2). And the number of farmworkers living in poverty is increasing ("Report"2).

Many of these workers face lack of health care, no pension plans, absence of paid vacations, and no overtime pay, unemployment compensation, or workers' compensation.[6] Less well known is that these workers have experienced a range of problems that result from their exposure to pesticides (Acury 2). Indeed, after mining, farming is the second most dangerous occupation in the United States ("Report" 3). This country uses more pesticides than any other in the world (Perfecto 179). In 1986, we used 1.2 billion pounds of pesticides, garnering 4.6 billion in sales; today we use even larger amounts. Farmworkers themselves are the most highly exposed to these chemicals because of the extensive hand labor that most perform and because they lack the power to influence farm safety (Acury; Aseplin; Gianessi; U.S. EPA, *Pesticides*). In the late 1980s, producers combined more than 600 active ingredients with "inert" ingredients, many of which have since been shown to have palpable effects, to form approximately 35,000 different chemical formulations designed to kill plants, insects, fungi, bacteria, nematodes, and rodents (U. S. EPA, *Pesticides*). Today, of the 85,000 different synthetic chemicals manufactured in the United States, more than 3,000 are produced in what we classify as "high volume," and more than 75 percent of this group "have undergone no screening for possible developmental effects on fetuses and children" (Steingraber *Having* 88). Numerous studies have suggested that we still do not understand the effects of such combinations on human beings, wildlife, and the environment at large. Less than 10 percent of the products currently being used have been tested for effects of exposures to single ingredients, not to mention that the myriad combinations of ingredients have differing effects, depending on their exact mixture (Colburn et al; Perfects 180; Schettler et. al.). Without a doubt, farmworkers and those involved with pesticide production suffer the most from the United States' chemical dependency on pesticides, and from the dependency of agribusiness on the export of such chemicals abroad. Estimates from 1985 suggested that around 313,000 farmworkers in the United States suffered from pesticide-related illnesses each year (Wasserstrom and Wiles). Today all of the estimated 4.2 million migrant and seasonal farmworkers remain at risk for environmental and occupational illness and injury, in addition to health disparities associated with poverty (Acury 2). A testament to the deplorable health conditions can be seen in the alarming figures that the life expectancy of farmworkers is forty-nine years (twenty-five years below the national average) and that the infant mortality rate is 125 percent higher than that of the general population ("Report" 3).

A 1993 study by the General Accounting Office (GAO) found that between 3.2 and 4 million farmworkers faced significant exposure to pesticides either directly through application activities, or indirectly through contact with residues on treated fields (Pesticide Action). While the U.S. Environmental Protection Agency (EPA), at that time, estimated at least 20,000 illnesses associated with the occupational use of pesticides on U.S. farms each year, other estimates have placed the numbers at 300,000 (PAN International). Today, the EPA estimates that approximately 300,000 farmworkers experience acute illnesses from pesticides each year (Gay 2). Some argue that all of the estimated 4.2 million migrant farmworkers are at risk for serious environmental illness (Acury 2). But it is important to recognize that such estimates are purely that—estimates. As with other environmental toxins in the United States, we do not have a good mechanism in place for tracking exposures and health effects. For example, in 1978, the EPA adopted a Pesticide Incident Monitoring System (PIMS) to coordinate and collect information on the adverse effects of pesticides, but Congress cut the funding for this program in 1981(Pesticide Action). Since then, the EPA has continued to receive voluntary reports of pesticide incidents and, in 1991, it set up the Incident Data System (IDS) that takes data from voluntary sources. As with many policy measures instituted in recent years, such reporting may appear responsible but does little to track exposures, given their voluntary nature. More recently, pressure by industry and conservative lawmakers have not boded well for efforts to systematically track environmental exposures. In particular, the Patriot Act closes down access to information about many environmental exposure sources.

Compounding this is the fact that in many places legislation is scant for protecting farmworkers from pesticide exposure. For example, figures for 1999 in North Carolina show that of the growers providing housing who submitted to water testing, 44 percent had contaminated water, while one washtub for every thirty workers met North Carolina's requirements and only 4 percent of farmworkers there had access to drinking water, toilets, and handwashing facilities in the fields (Smith-Nonini). Although we regulate other occupational exposures, we do not do so for pesticide exposures to farmworkers. Indeed, the laws that govern pesticides—the Federal Food, Drug and Cosmetic Act (FFDA) and the Federal Insecticide, Fungicide and Rodenticide Act (FIRFA)—set legal limits for pesticide residues allowable in foodstuffs. Levels of exposures to pesticides are based on so-called acceptable daily intake of exposures on foods, not on occupational exposures (Bullard 1993). It bears repeating that FIRFA requires companies manufacturing pesticides to test for toxicity, but until old, untested pesticides are measured, they can continue to be used. Steingraber comments that "as one critic noted, it is as if the bureau of motor vehicles issued everyone a driver's license but did not get around to giving us a road test until ten years later" (Living). In addition, "According to the National Research Council, only 10 percent of pesticides in common use have been adequately tested for hazards" (Living). When standards have been implemented, they have been based on studies of healthy men—without regard to other, more sensitive groups, incuding women, children, and babies (Institute; Gay). When we look at the situation outside our borders, in many cases the setup is even more alarming. For example, as late as 1994, ten pesticides that have been banned for use in the United States were still authorized for use in Mexico (Barry 1994, quoted in Gay). In addition, we continue to export products banned for use here, including over 4 million pounds of endocrine-disrupting pesticides that were shipped to Mexico in 1994 alone (FASE). Part of the persistence of the problem stems from the fact that World Bank and IMF structural adjustment policies "have been influential in promoting pesticide sales and in promoting agriculture for export, which is more likely to use pesticides," while NAFTA "encourages the expansion of agricultural exports which are 'characterized by heavy use of pesticides'" (Dinham; Thrupp—both quoted in Gay).

BREASTMILK CONTAMINATION AMONG U.S. FARMWORKERS

While little research exists about exposure to toxins for infants of farmworkers, the work that has been done on breastmilk contamination of other groups and on more general toxic exposure of agricultural farmworkers can give us an idea of the potential levels of exposures in farmworkers' babies. First, let's look at breastfeeding rates for these women. A study, "Breast-Feeding Intentions and Practice among Hispanic Mothers in Southern California," found that "the likelihood of intending breast-feeding was greater for mothers who migrated from Mexico than for mothers born in the United States" (odds ratio 4.75). These mothers also tended to carry through with their intentions; "the likelihood of breastfeeding practice was greater for mothers who . . . migrated from Mexico compared with mothers born in the United States (odds ratio 8.54) (Romero-Gwynn and Carias 626). In addition, this study found that, contrary to findings for other ethnic groups, among Latina women, the least educated were most likely to breastfeed (Romero-Gwynn and Carias 631). Another study, published in 1989, found that "infants in households for which the preferred interview language was Spanish were more likely to be breast-fed than were infants living in households for which the interview was conducted in English" (John and Martorell 868). The high proportion of Latino farmworkers, along with the fact that these workers fit the profile for those most likely to nurse in the "Breastfeeding Intentions Study," suggest the likelihood that a proportionately large number of agricultural farmworkers do nurse. One of the few studies on Mexican agricultural migrant workers in the United States found that 83 percent of the respondents breastfed, although the researchers note that the majority of their respondents who breastfed "did not rely on breastmilk as the sole source of nourishment for their infants" (de la Torre and Rush 735).

Sandra Steingraber points out that, of the pesticides and exposures occurring in vitro, "Those with low molecular weights cross the placenta without restriction," but that pesticides "made of bigger, heavier molecules are partly metabolized by the placenta's enzymes before they pass through, but sometimes this transformation makes them *more* toxic, placing the fetus at even greater risk" (*Having* 34). In addition, she notes that "only one study of environmental contaminants in amniotic fluid has ever been done," and

that it found "detectable levels of organochlorine pesticides in one third of
the thirty samples of amniotic fluid tested" (75). The research subjects in
this study were not farmworkers or women more heavily exposed to pesticides than the general population—which suggests even higher exposure
rates for farmworkers. What is alarming is that "most chemicals have not
been tested for their ability to have teratogenic effects" (effects to the fetus)
or effects brought about through breastmilk contamination (Steingrabber
Having 88). When we look specifically at agricultural chemicals and their
effects on infants and children, we see that those that came on the market
before testing for fetal toxicity was required are either exempt from analysis
or still awaiting further evaluation, and so are still sold and used (*Having* 89).
A Johns Hopkins Report noted that "some pesticides currently being used
may be developmental toxicants" (Pew 35).[7] A study published in *Environmental Health Perspectives* in 2002, which evaluates the "take-home" pesticide
exposure among agricultural workers and their children, found that, in looking at house dust and vehicle dust samples, that "the take-home exposure
pathway contributes to residential pesticide contamination in agricultural
homes where young children are present." Although the study did not look
at breastmilk contamination, one might extrapolate that residues brought
home would affect the breastmilk of mothers (Curl, Fenske, et al.).

BUYING ORGANIC AND PROTECTING "OUR" CHILDREN: HOW WE FORGET THE "OTHERS"

We often talk about limiting our exposure to pesticides by buying organic.
But when we consider the effects of pesticide use, we must realize that not
only consumers are at risk. Clearly, one action we, as consumers, can take
to reduce our exposure to pesticides is to "buy organic." As we discuss how
accessible or inaccessible organic products are to various population groups,
I sometimes ask my students to think about the marketing process that
allows a grocery chain where we live to make organic foods available to those
who shop at its suburban stores, but denies such choices to those who live
in the inner city and shop at the "ghetto chopper." We discuss how to sort
out claims that the grocery chain simply offers what sells in a particular area,
and weigh these against observations that racism and classism play a role in
such practices, whether they get justified as "sound business practices" or not.
But if we move beyond questions about the impacts of pesticide use on
various consumer groups, to questions about the impacts on farmworkers or
on workers in pesticide production, or on those "downstream" from both, a
more complicated picture emerges. Although the health of workers is utterly
significant and rarely addressed, it is not just the workers' health that must
be taken into account.

    In an article that traces the multiple effects of pesticide use on
farmworkers, Ivette Perfecto offers an example of what happened when

Nicaraguan agronomists in the 1950s decided to use pesticides on cotton fields. Aside from the health effects of the pesticides, which can be significant, an effect not often recognized involves how that action has impacted later generations. Perfecto asks "When the cotton fields of northern Nicaragua were poisoned with pesticides in the 1950s, what was expropriated from whom?" (178). The answer? Farmers of later generations trying to grow soybeans in the fields where cotton had been planted find that the beet armyworm (a caterpillar pest of soybeans) makes the task impossible (178). In the vein of Carson, Perfecto points out that the beet armyworm is not a "natural" pest, but rather that "it was created as a consequence of the spraying of pesticides in the cotton fields of the 1950s." If the pest had not been "created" by the decision of agronomists in the 1950s, later generations of Nicaraguans could benefit from soybean production. Perfecto wonders "Exactly how much are they losing by not being able to produce soybeans?" Observing that "many other examples could be cited," she answers that "that quantity of loss represents a quantity that had been expropriated from them by the cotton farmers of the 1950s."

Writing in 1992, as the environmental justice movement just began to surge into power, she notes that "the point is simply that environmental degradation can be represented, so as to be parallel with class exploitation and the exploitation of people of other 'races,' as the expropriation of someone else's potential, and that someone else can be thought of as people in future generations" (178). The connection needs to be reiterated, and this example bears it out—systemic structures determine that the most vulnerable often bear the consequences of decisions, whether conscious or not, of those in positions of power. Perfecto states that "people belonging to a 'race' that is considered lower, have no right to resist, nor do unborn generations have the political power to resist" (179).

Perfecto moves on to discuss a situation involving pesticide use in the United States and the ways in which, in the early years of environmental awareness, pressure by environmentalists to protect consumers, particularly infants and children, from pesticide effects resulted in greater degrees of exposures to farmworkers and their infants and children. When addressing the impacts of pesticide exposure in general on Latino and other farmworkers, including impacts from breastmilk contamination, one important part of the picture involves the recognition that health problems faced by these populations get exacerbated by other issues, including the effects of poverty.

The health effects of pesticides are difficult to document for several reasons, including the dearth of studies on cumulative and combination doses, the lack of information on exact exposure rates, and the varied but sometimes lengthy lag time between exposure and health effects. Researchers also make broad distinctions between acute and chronic effects. Perfecto explains that acute effects range from dizziness, to vomiting, to eye and skin irritations, to respiratory problems, to more systemic effects including death, while chronic effects include a range of cancers, reproductive malfunctions, a broad range of developmental and behavioral growth problems, and birth defects (181). She also notes: "These residues tend to accumulate in animal fatty tissue and reach extremely dangerous levels" so that "breastmilk, which has a high concentration of fat, has been found to be contaminated with a variety of pesticides in women in the rural South" (181). Key to her discussion is the fact that, in general, pesticides that persist in the environment tend to have lower acute toxicity, while those that degrade more rapidly tend to be more acutely toxic (181).

While the general contamination of farmworkers, and that of their infants and children through breastmilk exposure are what interest me here, Perfecto offers an example of a situation in which types of exposures were impacted by pressure lodged by mainstream environmental groups. This example, from the period immediately preceding the dawn of the environmental justice movement, reminds us of the importance of the work of such groups, and offers an illustration of one instance of environmental racism impacting the lives of Latino farmworkers. According to Perfecto, as environmentalists became aware of toxics issues due to pesticide exposure, "although unconsciously, a decision was made to protect the environment (and pretty animals) at the expense of the health of farm workers" (182). In this particular case, what happened was that with the banning or restriction of less toxic but highly persistent organochlorines, such as DDT, aldrine, chlordane and others, farmers began to use less persistent but more acutely toxic pesticides such as N-methyl carbamates and organophosphates, such as parathion. In other words, "high acute toxicity correlates with rapid degradation and low acute toxicity correlates with persistence." Perfecto points out that "this fact set up a natural confrontation between those who work directly with pesticides (minority farm workers), and those who experience residues and secondary effects only (the general population as consumers and as future generations)" (201). She notes that "the decision was effectively made by those in power to protect the environment (for the enjoyment of future white generations), even if it meant the further exploitation of farm workers (in other words, people of color)." In a surprising twist, Perfecto "blames" Rachel Carson, claiming that the "switch came as a consequence of the environmental awareness in the U.S., instigated largely as a response to the book by Rachel Carson, *Silent Spring*." She points to how environmental concerns and EJ concerns can sometimes be at odds. Musing how a "class-free, racism-free society" would have demonstrated concern over both the long-term and short-term effects—the health of consumers and their offspring, as well as that of farmworkers and their offspring—Perfecto blames Carson, rather than the short-sightedness of some environmentalists. This example reminds us of the need to maintain a broad view, rather than to go with knee-jerk reactions.

What Carson and many later environmentalists have lamented and fought against is that the pesticide industry has worked to prevent the development of nonchemical pest control, while systemic structures of power continue to allow such practices to operate unchecked. Getting at part of the problem, Perfecto blames the "structure of capitalism," noting that it "promotes whatever expropriation will provide that 'all-important' competitive edge necessary for capitalist accumulation" (182).

MOVING TOWARD ENVIRONMENTAL JUSTICE

We now know that after an initial period of resistance, many in the more mainstream environmental movement have begun to understand the claims of environmental racism and injustice that EJ activists have lodged against them. And just as many activists have been working to address environmental injustices, many activists have been working to remedy some of the injustices faced by farmworkers and by the increasingly Latino population that now makes up the largest "racial/ethnic" group in the United States. An example of the kinds of work being done to address issues of farmworker safety includes a study conducted in 1999 with 293 farmworkers in eastern North Carolina as part of the Preventing Agricultural Chemical Exposure in North Carolina Farmworkers' Project. Noting that "efforts to provide safety training for farmworkers have not been fully evaluated," and that evaluations of the "most comprehensive pesticide safety regulations for all agricultural workers"—the U.S. EPA's Worker Protection Standard—have not included any direct data collection with farmworkers, the study examines "how safety information affects perceived pesticide safety risk and control among farmworkers and how perceived risk and control affect farmworker knowledge and safety behavior" (Acury 2).

Stating that "a key tenet of environmental justice is that communities must have control over their environment," the authors of the study looked at issues of perceived risk and perceived control, such as whether workers "believe their health is hurt by pesticides," "launder work clothes separately," "shower after direct contact with chemicals," or "wash before going to the bathroom." They conclude that the results "argue that for pesticide safety education to be effective, it must address issues of farmworker control in implementing workplace pesticide safety." A factor at stake is that some farmworkers expressed little concern about the risk of pesticides hurting farmworker children (26 percent saw this as of little or no concern) or unborn children (28 percent). This could translate into lack of awareness regarding the possible contamination of breastmilk or of actions that could be taken to protect breastmilk from further contamination. A study by Vaughan, looking at California farmworkers, found that these workers perceived little control over their exposure to chemicals and their negative health effects. The ability of farmworkers "to communicate with their employer," negotiate "power relationships at work," and to have access to protective equipment are all important factors for their safety that the study documented. It also suggested that there is work to be done on improving relations between workers and employers and in addressing the feeling among many farmworkers that they do not have power to negotiate change.

While I have not been able to locate work on breastmilk contamination in Latina women as it relates to pesticide issues, a group of scientists working for the Washington State Department of Health have published an article on "The Effects of Fish Consumption on DDT and DDE levels in Breast Milk Among Hispanic Immigrants." Results indicate that "fish consumption did not significantly increase DDT/DDE in breast milk concentrations." But, it did find that subjects born in Mexico had elevated levels of DDT/DDE in breast milk compared to levels found in U.S. born subjects regardless of fish consumption." When the researchers compared infant daily intake levels for the various subject groups, including those with the higher contamination levels, they determined that, at the current time, "breastfeeding should still strongly be recommended."

Considering "Chicana Strategies for Success and Survival," Kamala Platt points out that "in the last decade, Chicanas have been prominent in a growing environmental justice movement that links environmental injustice with structural racism and patriarchy, identifies environmental racism as the outcome of colonialism and imperialist capitalism, and critiques 'mainstream environmentalism'" (48). Indeed, she notes that "through a diverse metaphorical iconography utilized to promote environmental justice Chicana poetics address issues that include farm worker organizations' fights against agrobusiness pesticide misuse, environmentally related health concerns about women working in and/or living near the U.S.–Mexico border *maquiladoras*, and the amelioration of the toxic waste dumps, all of which disproportionately affect communities of color" (49). According to Platt, what is at stake for these groups is "expanding the agendas of the environmental and social justice organizations that preceded them," which involves articulating the "the relevance of the interconnections between 'who they are' and 'what they want'" (50). Part of this "who they are," according to Platt, involves the recognition that for many in environmental justice communities, this means gender, in addition to race, ethnicity, and class issues. Indeed, for her, many of these activists see "issues of class and gender" as "vital factors that must be studied in order to strategically resist (what is most often) corporate-or government-sponsored ecocide, genocide, and destruction of communities, livelihoods, cultural traditions, and personal health" (52).

Mary Pardo, who studies Latina activists, asserts "that through much sociological writing, women, ethnic/racial minority group members, and working-class men and women [are] often either absent from the discourse or bec[o]me the victims of social problems rather than active participants in social relations" ("Identity" 54). While we see strong self-representation in the Mother's Milk Project, its literature, and its presence as a force in the story, specifically, about the environmental contamination of breastmilk, and while we witness a strong self-representation on the part of Chicana activists working on other toxics issues, I have not been able to locate attention among Chicana or farmworker activists to the issue of breastmilk contamination. Platt points out that one of the Mothers of East L.A.'s flyers boasts that "[T]he Mothers have also marched with Cesar Chavez to protest pesticides, lobbied for the Mexican preservation of Olvera Street, assisted in Voter Registration and Citizenship drives and partic[i]pated in numerous grassroot[s] political campaigns," including building alliances with other Chicano groups such as the United Farm Workers and with other environmental groups such as Greenpeace (Platt 56). Work is being done, and will continue. But raising awareness of the likelihood of breastmilk exposures can only bolster the resolve in women in these groups to work in coalition with others to bring an end to practices contributing to such exposures.

URBAN WOMEN AND THOSE AT RISK DUE TO OCCUPATIONAL OR OTHER EXPOSURES

When we consider other groups of women at risk for high levels of environmental pollutants in breastmilk, we can identify several as being more vulnerable than the general population. These include women eating large amounts of locally caught fish in areas where fish has high levels of contaminants, women living in areas with higher concentrations of pollutants than the general population, older mothers (because their body burden of toxic chemicals has had a longer period in which to accumulate), and women subject to occupational exposures, either from their own work or because their husbands or domestic partners bring residues home on clothing and shoes. While I have not located any studies of breastmilk contaminants in women living near landfills, a study published in the British medical journal *Lancet* of people living up to three miles from hazardous waste landfill sites found that that these women have a 40 percent higher risk of giving birth to babies with congenital chromosomal abnormalities, such as Down's Syndrome, than those who live further away (Dolk, Vrijheid, Armstron, et al.).[8] This suggests that such women might also be at risk for higher levels of contaminants in breastmilk.

Researchers have looked at dry-cleaning workers and residential exposure near dry cleaners for information about possible exposures to infants breastfed by these groups. One study reported obstructive jaundice and hepatomegaly in a six-week-old breastfed infant exposed to PCE when his mother regularly visited his father who worked in a dry-cleaning factory (Schreiber "Parents"). This study showed that the "cessation of breastmilk exposure resulted in rapid improvement of the child's condition" (Schreiber "Predicted" 517). Dr. Judith Schreiber, who currently works for the New

York State Attorney General's Office and has published an extensive disser-
tation and other work on the environmental contamination of breastmilk,
notes that "no breastmilk sampling has been reported of women who currently or previously were employed in the dry-cleaning industry, or of women
residing in proximity to dry-cleaning establishments" (518). But she surmises
"[t]he elevated airborne levels of PCE [perchloroethylene, or "perc"] typically
found in dry-cleaning establishments and nearby residences, coupled with the
relatively long adipose tissue storage of absorbed PCE, point to the likelihood of
finding significant concentrations of PCE in the milk of exposed women" (518).
She says that "exposure to PCE is of concern due to its toxicological properties
demonstrated in humans and animals," pointing out that "most of the modeled
exposures" in a study looking at seven scenarios of exposures of mothers working
or living near dry cleaners "result in infant milk doses which exceed the RfD
[oral reference dose of 0.01 mg/kg/day], in some cases by several orders of magnitude" (520). While Schreiber notes elsewhere that, given its superiority over
formula in many capacities, breastfeeding is advisable for all women except those
most highly exposed, this appears to be one of the subgroups where questions
remain. Noting "the predicted levels of PCE in breastmilk under several of the
exposure scenarios evaluated suggest that an infant may be exposed to doses of
PCE within an order of magnitude of doses associated with adverse health effects," she urges researchers to engage in monitoring studies of women expected
to have high exposure rates from dry cleaners "so that appropriate risk management alternatives can be better evaluated" (523).

APPROACHES TO REMEDIATION: COMMUNITY-BASED
PARTICIPATORY ACTION RESEARCH

A 2002 paper published as part of a series on "Advancing Environmental
Justice Through Community-Based Participatory Action Research," a supplement of *Environmental Health Perspectives*, looks at a case of subsistence fishing hazards in the Greenpoint/Williamsburg section of Brooklyn, New York.
While the work done there did not specifically address breastmilk toxicity,
it did examine community health as it relates to environmental pollution
from subsistence fishing, widely recognized as a key source of breastmilk
contamination. I turn to this project because it offers insight into the contributions of the emerging field of participatory action research and suggests
how some of the EJ work impacting breastmilk toxicity levels has played out.
    First I want to discuss the broader issue of community-based participa-
tory action research (CBPR). In CBPR, scientists, social scientists, and other
researchers "work in close collaboration with community partners involved
in all phases of the research, from the inception of the research questions
and study design, to the collection of the data, monitoring of ethical concerns and interpretation of the study results" (Shepard, Northridge, Prakash,
and Stover 140). This model, in contrast to earlier, hierarchically based

methodologies in which researchers studied communities without consulting members, hearing their concerns, or explaining research result to them, accompanies the assumption that community members are "experts" and have knowledge that is important to solving problems. In CBPR, researchers discuss findings with broader communities, including residents, media sources and policymakers, all aimed at impacting policies and practices so as to improve overall health and well-being. In addition, CBPR works to "build capacity and resources" in communities and to facilitate government agencies and academic institutions to better understand and incorporate community concerns into their research agendas (Shepard et. al 140). The holistic, integrative and unifying strategies characteristic of CBPR that address multiple areas simultaneously are especially important in EJ contexts.

In a six-year collaboration between West Harlem Environmental Action (WE ACT), an environmental justice organization; the Harlem Health Promotion Center (Harlem HPC), an academic center dedicated to advancing the science and scholarship of CBPR; and the National Institute of Environmental Health Sciences (NIEHS) Center for Environmental Health in Northern Manhattan at the Mailman School of Public Health, several projects were begun, including air monitoring studies, training courses for community leaders, educational forums for community residents, and lobbying efforts (Shepard et al. 140). One of those endeavors, which involves the Watchperson Project, a community-based organization in the Greenpoint/ Williamsburg neighborhood of Brooklyn, illustrates a key tenet of the environmental justice movement. It challenges researchers and decision makers "to acknowledge that scientific expertise is necessary but insufficient to address the multiple and persistent hazards facing the poor and people of color" (Corburn 242). Because people of color and those disadvantaged economically continue to bear a disproportionate burden of the results of our policies and practices, and because these groups often face multiple exposures that are often overlooked by mainstream agencies and policymakers, many underprivileged communities have begun to demand a participatory role in defining, analyzing, and creating solutions to environmental problems. The Watchperson Project illustrates the ways in which valuing local knowledge, contextual thinking, and experiential information-gathering play key roles in developing vital new ways of processing and addressing problems that are often difficult to approach, given their systemic nature.

THE WATCHPERSON PROJECT IN THE GREENPOINT/ WILLIAMSBURG NEIGHBORHOOD OF BROOKLYN, NY: BRINGING AWARENESS TO SUBSISTENCE FISHING HAZARDS

Approximately 160,000 residents live in the Greenpoint/Williamsburg neighborhood, an area spanning less than five square miles (Corburn 243). Among the poorest areas in New York City, 35.7 percent of its population lives below the poverty line; 42 percent of the residents are Latino (mostly Puerto Rican and Dominican), 24 percent are Hasidic Jew, 13 percent are African American, and 10 percent are Polish and Slavic immigrants (243). The "environmental profile" of the area indicates a problematic legacy; for example, it boasts the largest amount of land—12 percent—used for industrial purposes than any other of New York City's fifty-nine communities (243). In addition, it houses a disproportionate number of polluting facilities, including a sewage treatment plant, thirty solid waste transfer stations, a radioactive waste storage facility, thirty facilities that store extremely hazardous wastes, seventeen petroleum and natural gas facilities, and ninety-six aboveground storage tanks (243). To compound this, only 3 percent of the neighborhood is shaded by trees, compared with 11.4 percent tree coverage for Brooklyn and 16.6 percent for New York City (243). While little has been done to document health effects from these burdens, two New York City Department of Environmental Protection-supported studies found that incidences of stomach cancer, certain types of leukemia in men, pancreatic cancer in women, cancers of the central nervous system and certain leukemias in children are among the highest in New York City (243 footnote 18, 19).

The CBPR project in this area demonstrates two advantages of such community-based work; first, it suggests how such projects can help mobilize residents to take action, and, second, it illustrates how community involvement can become a vital part of conventional scientific work (247). As part of the Cumulative Exposures and Subsistence Fishing project, the Community Exposure Project (CEP) constitutes an attempt by the EPA to move beyond the "single-source, single-hazard approach to assessing risks" that has characterized much environmental work. It involves the EPA in a partnership with the Watchperson Project and together setting out with the recognition that people of color, the poor, and other sensitive populations face simultaneous exposures to multiple environmental pollutants from multiple sources, often at higher rates than the general population. Cumulative assessments differ from "traditional single-source risk assessments" in that they "consider multiple pathways, sources, and endpoints" and focus on "populations, not individuals, and aggregate . . . by population subgroups, such as those highly exposed and highly sensitive," a factor important in considering the environmental contamination of breastmilk (Corburn 247). This also divides highly sensitive populations including those with preexisting conditions, age (i.e., infants), and gender (i.e., pregnant women) and recognizes multiple pathways and potential routes of exposure, including bioaccumulation and biomagnification, which are also extremely important in thinking about environmental exposure through breastmilk (247).

Samara Swanston, director of the Watchperson Project, suggests that the EPA methods for assessing dietary intake seemed problematic to many residents, given that they are based on assumptions that lack specific under-standing of the diets of the diverse ethnic populations and ignore the poten-tial exposures from eating locally caught fish. In the course of community meetings, residents pointed out to EPA officials that many of them were subsisting on a diet heavy in East River fish; EPA representatives noted that they had not heard of this potential health hazard before. Because many anglers are immigrants and nonnative speakers, possibly reluctant to share information with outside researchers, this seemed a viable opportunity for a community-based participatory project. Community members pointed out that in order to obtain angler exposure data, local people would be the appropriate avenues for gathering information, as local anglers would be more likely trust these information-gatherers. According to one EPA official, "we had no choice but to let the community groups gather the data, for a number of reasons, including language, cultural barriers and potential trust issues, we felt the local people could best gather this data" (248). Addressing one of many benefits of CBPR, he notes that this was one situation "where residents raised an issue we hadn't considered, defined the extent of the problem, and provided data for analysis" (248).

The Watchperson Project worked with the EPA to develop interview protocol that involved types of fish consumed, frequency and amounts of consumption, as well as racial, ethnic, and demographic characteristics of the anglers and their families. Many of these anglers rely on the East River fish as a major source of food, with responses such as "Look, this is my way of feeding my family. I ain't got no job and this is what I did in the D.R. [Dominican Republic]. I got to feed 5 or 6 people a night. Know what I mean?" being typical (248). The project allowed EPA representatives to determine that several contaminants, including cadmium, mercury, chlor-dane, DDT, dieldrin, dioxins, polychlorinated biphenyls, arsenic, and lead were being ingested by the anglers and their families in significant dosages. Indeed, the estimated exposure levels for the anglers and their families "ex-ceeded U.S. EPA oral reference doses which generally serve as benchmark levels for noncancer health effects for all contaminants except cadmium at both low- and high-end consumption estimates," with exposure to dioxins being particularly high (248). EPA's calculated lifetime cancer risk for adult subsistence anglers was found to exceed 1 in 10,000, which is significant given that its acceptable cancer risk is 1 in 1 million. This case is important, especially since without the community input, the EPA would have over-looked this health hazard. It is also significant that, as far as I have found, no attention was given in this case, or in similar ones looking at exposures to anglers, to likely exposures that we would expect the breastfed infants of angler mothers to be accruing.

The case also demonstrates that people getting together can make a difference. The Watchperson Project, realizing that the EPA would take more than five years to complete the community exposure profile for the neighborhood, began a series of "fish-ins" to educate anglers about potential toxic contamination, and to work on cleaning up the riverbank. Community members created materials, printed in Spanish and English, that described possible health risks, and educated the people about techniques that could reduce their intake of contaminants. They also worked on developing alternative foodsources such as community gardens and shared their resources to help the New York State Department of Environmental Conservation to develop culturally sensitive practices for advertising and enforcing fish advisories (249). A final note about CBPR involves the recognition that one of the limits of such collaborative work is that it cannot go far enough in addressing the structural issues that make for inequitable distributions of the hazards of environmental practices and policies (255). While it is a start, more needs to be done.

From time to time, other groups of researchers have investigated the fishing habits and possible ingestion of toxins through contaminated fish on the part of anglers (Foran et al., Humphrey). And some have focused on anglers of color in certain regions. For example, noting that "concerns about toxic chemicals in Michigan surface waters" have been an issue, a team of scientists—Patrick West, Mark Fly, Frances Larkin, and Robert Marans—looked at minority anglers and toxic fish consumption in conducting a statewide survey of Michigan, since in 1991 and 1992 "pilot research by West indicated that black fishermen consumed more than three species of fish from the Detroit river than white fishermen" (West 100). Finding that fishermen of color and their families consumed 21.7 grams per person per day as opposed to 17.9 grams for whites, with 20.3 grams for blacks, 24.3 for Native Americans, 19.8 for "others" including Latinos, other racial and ethnic groups, and those identifying as being of "mixed" descent, the study suggests that the infants of these populations would be more at risk for exposures than the infants of white anglers. In addition, finding that "there were numerous minority sub-groups in the sample who had fish consumption rates exceeding 30.3 grams/person/day," and that these figures are "almost five times as high as the 6.5 gram assumption in Rule 1057," the law then governing the State of Michigan's regulation that controls the discharge of toxic chemicals into Michigan surface waters, the study suggests that to protect such populations, regulations need to take into account the groups most likely to consume affected fish, as well as the most sensitive among those populations—the nursing infants of these mothers. As we can see from the more recent work done in the Greenpoint/Williamsburg neighborhood, policymakers are only now beginning to take the consumption habits of anglers, especially anglers of color, into consideration. And even more troubling is the fact that breastmilk contamination never entered the discussion of that progressive study.

GREATER MOVEMENT TOWARD ENVIRONMENTAL JUSTICE:
THE INSTITUTE OF HEALTH'S VISION AND THE LA DUKE, BONNIE
RAITT, AND INDIGO GIRLS CONSCIOUSNESS-RAISING TEAM

Recognizing that environmental injustice constitutes one of many outgrowths of a long history of injustices, including the effects of the system of slavery and the racial and ethnic oppression that followed, the results of waves of colonization, of the genocide of Native Americans, and of multiple forms of oppression that continue to haunt us, those who want to end such injustice must realize that positive change will require action from many angles, along with patience, perseverance, and a willingness on the part of those more privileged with power and opportunity to understand that privilege and use it to work toward justice. One approach to positive change has come out of the recognition on the part of analysts that the division between "the environment" and public health that was instituted starting in the 1970s, has not been a healthy one (Wiant). In an effort to address this problem, the Institute of Medicine (IOM) conducted a workshop entitled "Rebuilding the Unity of Health and the Environment: A New Vision of Environmental Health for the 21st Century." Held on June 20–21, 2000, the workshop began by acknowledging that "the goals of environmental health are to maintain a healthy, livable environment for humans and other living species—an environment that promotes well-being and a high quality of mental and physical health for its inhabitants" (Institutde of Medicine 142).

Acknowledging that addressing environmental health problems requires "thinking about environmental health on multiple levels," and merging "various strategies to protect both the environment and health" simultaneously, workshop organizers sought to "raise awareness, promote community-based environmental health, and mold multidisciplinary partnerships to redefine and improve environmental health" (Lee 141,2). Important to the question of breastmilk toxicity is that organizers began by noting that "some subsets of the population are inherently more susceptible to cellular or genetic damage for a number of reasons, including genetic susceptibility, nutritional status, other social or cultural factors, or in the case of children, the vulnerability of developing systems to environmental insult" (142). Although the workshop did not address breastmilk contamination specifically, it offers the kind of approach that can help bring us where we need to be on this issue. For example, rather than being a workshop conducted by insiders for insiders, it involved a broad group of representatives, including "business leaders; economists; architects; urban planners; engineers; public health [officials]; environmental[ists] and social scientists; clergy; educators; and citizens" coming together to share and to discuss their views on the elements for a healthy environment (142). The unifying strategy became the necessity "to merge various strategies to protect both the environment and health" (142).

Robert Bullard, a leader in multiple EJ endeavors, emphasized at the workshop that "health is more than the absence of disease," that "environmental justice must be the starting point for achieving healthy people, homes, and communities," and argued for moving away from dichotomizing models such as "jobs versus the environment or jobs versus health" (143). The Interagency Working Group on Environmental Justice (IWG), which Clinton established under Executive Order 12898, has sought to develop collaborative models that bring together federal agencies with local groups. Under its auspices, fifteen environmental justice demonstration projects in diverse geographies have been initiated. The projects link "environmental cleanup, economic revitalization, and holistic community planning" in order to build community power and combat environmental distress. For example, one target community in the Arkwright and Forest Park sections of Spartanburg, South Carolina, a predominantly African American community located "within a quarter mile of two Superfund sites and close to an abandoned textile mill, an operating chemical plant, two waste disposal sites, and several suspected illegal dumps," is being revitalized with the help of a community-based organization there called Re-Genesis (144).

This group has been working in conjunction with a partnership of community groups, local government, business and industry representatives, faith groups, university members, and federal agencies to clean up and create a vision of revitalization that will work for the community. The participants envision new housing, technology and job training centers, along with a greenway to promote physical and emotional wellness and serve as a tool for combating obeisity in the community. The collaboration also involves establishing a health clinic, which might serve to promote breastfeeding and awareness of possible environmental toxins in breastmilk, while also educating on how to limit these exposures. This is important, given that while rates of breastfeeding among African American women remain low in comparison to other groups, they are steadily increasing. While such projects will take time, it is important that workshops such as that conducted by the IOM are being held and serve as workable models for future projects. The IOM organizers note that such work involves long-term and short-term goals, including educating our national leaders about environmental justice issues, promoting a national dialogue about the ramifications of inaction, fostering "capacity building" within communities so that those affected can form strategic partnerships, and identifying elements of successful collaborative models so that they can become tools encouraging further remediation work (145).

Another example of a creative strategy for raising awareness of breastmilk contamination involved a collaborative effort between Winona La Duke, Bonnie Raitt, and the Indigo Girls. Banding together to bring attention to Katsi Cook, the Mohawk Mother's Milk project, the environmental contamination of breastmilk, and the broader issue of environmental injustice,

the popular speaker and singers went on tour. Their efforts raised the awareness of thousands of listeners who attended their concerts. They also drew in
the listening ear of those in the position to bring about policy changes. La
Duke comments on her own participation:

> I discovered that if you take Bonnie Raitt and the Indigo Girls with
> you, you can get a meeting with Carol Browner at the EPA. So that's
> what we did. We got Gudgy Cook a meeting with Carol Browner . . . we
> got Carol Browner to pay attention to the people of Akwasasne for a
> time. . . . That's what we tried to do on that tour, draw attention to
> these issues and have women like yourselves relate to Gudgy Cook,
> relate and understand that she's a native woman, but she's a woman,
> and what she is facing is what you're facing. (Native 3)

Reflecting on the situation at another time, La Duke emphasizes what is
perhaps the most provocative point about the message delivered by Katsi
Cook and describes Cook's meeting with Browner: "She spoke mother to
mother, noting that the Mohawk mothers needed the EPA mother to help
them." Reminding us that we all bear this same connection, our early connection, however tenuous, to our own mothers, she notes, "The Mohawks
are hoping that the Great White Mother, The Environmental Protection
Agency, will do her job. That she will protect the water, the air, the soil, and
the unborn Mohawks" (All Our 23).

Krauss points out that the voices of white working-class, African
American, and Native American women activists "show the ways in which
their traditional role as mothers becomes a resource for their resistance"
(1993, 247). But motherhood as a unifying force is not all that is operating
in her article "Women and Toxic Waste Protests." These women's "emerging
analysis of environmental justice is mediated by differences of class, race, and
ethnicity." But this is not a negative—it can be a force for powerful change.
One of the hard lessons learned by feminists in the past several generations
is that women's perspectives are informed by their positionality. As Audrie
Lorde reminds us, "Certainly there are very real differences between us of
race, age, and sex. But it is not those differences between us that are separating us" ("Age" 115). Rather, it is "our refusal to recognize those differences, and to examine the distortions which result from our misnaming them
and their effects upon human behavior and expectations" (115). Exploring
cross-dialogue and critiques of certain groups by others are important for
allowing a range of voices to speak. What the La Duke/Indigo Girls, and
Bonnie Raitt performances did was to bring together many people from many
different perspectives in community gatherings. They offered education and
celebration, a model for the future that helps people to become aware of
environmental injustice, but also encourages the hope that, together, we can
forge positive change.

CONCLUSION

A NEED FOR MORE ATTENTION, AND MORE CAREFUL ATTENTION TO BREASTMILK TOXICITY

I began this book with a story about a missed opportunity, specifically about how our institutions have failed to support one pro-breastfeeding ad campaign. That story is part of a larger narrative, one that concerns the environmental contamination of breastmilk. When we confront the reality that the breastmilk of mothers in the United States contains a host of environmental toxins, it is important to recognize that a key part of this story involves information being systematically withheld, along with the systematic downplaying of breastmilk. The public will never see those ads asking us to confront the risks of formula-use. Many of us will also never realize that our institutions have worked together to allow our environment, and consequently our breastmilk, to be contaminated. I wrote this book because I believe that we can link the silencing of those advertisements to a much larger silencing that involves the degradation of our environment, our bodies, and our breastmilk. I noted how this institutional depriveleging of breastmilk in favor of industry pressure points to a larger set of institutional issues, and a history of supremacist assumptions and practices. Sexism, racism, classism, ageism, colonialism, corporatization—the environmental contamination of breastmilk tells a story about how structures of power have operated and continue to operate in such a way that some people's health and opportunities and, quite literally, their lives, get deprivileged, negatively impacted, or sacrificed. The less privileged among us—people of color, particularly Native Americans and Latino farmworkers, those working in certain occupations, those living in certain places, and those with less economic means—find themselves more subjected to environmental contaminants than others in the population. Shining a light on breastmilk toxicity means that

167

our babies, our future generations, must be added to these groups whose lives, health, and well-being are at risk.

I have alluded to the fact that, not so long ago, Sweden had breastfeeding rates comparable to those in the United States, and that, after several years of shifts in habits and policies, the vast majority of women in Sweden now nurse, most of them continuing past their babies' first birthday. Sweden provides hope that we can make such changes in our own lives. Sweden also offers lessons about the presence of environmental pollutants in human milk. After research showed very high levels of several contaminants in the breastmilk of Swedish women, people there lobbied for stricter environmental laws, so that today the breastmilk of Swedish women is much safer than in the past, and much safer than in the United States. We can bring about change; but it requires a shift in consciousness, in practices, and in policies. For such shifts to occur, we must confront our past, as well as our present, and we must choose to prioritize change. While change on the individual level is required, changes at the institutional level must be part of this process; without such changes women will not have the support they need to make nursing viable in today's work environment. And without such changes, women will not be able to continue indefinitely to use breastmilk with the assurance that the levels of carcinogens in their milk makes breastfeeding the better option. While some individuals have the privilege of making healthier lifestyle choices, such as eating organic, no individual, acting alone, can institute the policy shifts that will be needed to protect our world from unsafe pollutants. These require large-scale reprioritizing reflected at the level of public policy.

As I have described in earlier chapters, the combination of three occurrences has brought us to our current impasse. A historical situation resulting in a culture of resistance to breastfeeding has meant that we do not value nursing as a vital part of life or as an institution. A culture of immature ecological awareness has meant that we have not yet made the shifts that will need to occur for us to no longer promote "economic" development over health and well-being. And the continuing operation of racism, classism, colonialism, sexism, ageism, and corporatization has meant that supremacist gestures continue, leaving the less privileged vulnerable and determining that they have to work harder than others to have their voices heard and their rights upheld. If we are to become more responsible in our approach to infant nourishment, we will need to confront the ways in which many of the larger supremacist assumptions that inform our practices, our policies, and our gestures emerge, sometimes in hidden ways, in our approaches to infant feeding. These practices do not occur in a vacuum; they involve chains of events, perceptions, and habits that we inherit from the generations that preceded us and whose lives shape ours. They also involve choices that we make on a daily basis. And, most important, they have a significant impact on the legacy that we leave to future generations.

I noted how child sustenance becomes a figure for the oppressive frameworks at work in any historical juncture. In portraying how some women's milk has been perceived to be unfit or how all women' milk, starting in the 1930s, came to be seen as inferior to the man-made substitute, the story of wet-nursing reminds us of how sexism, racism, classism, and other oppressive assumptions have been inscribed in infant feeding practices in the past. Coming to terms with the suppression of breastmilk involves coming to terms with the most urgent social justice issues of our time. While wet-nursing narratives reveal a broad set of social, economic, and political upheavals shaking public confidence at particular periods, which often got articulated as fears about the transmission of disease, ill temper, or "weak nerves," the environmental pollution of breastmilk makes palpable the ways that some women's bodies become repositories for industrial toxins, while simultaneously are seen as somehow "polluted," or unclean, and so not worthy of protection. In other words, fears about rapidly shifting demographics (including the aftershocks of multiple colonizing ventures, increasing urbanization and industrialization, and anxiety over shifting social practices, work patterns, and family structures) became manifest in discourses about the dangers of wet nurses, which hid anti-immigrant and racist sentiment, class warfare, worries about job loss, and more general fears of social change. Today, the silencing of breastmilk's importance masks the silencing of a range of social justice issues. Classism and racism have emerged in concern over the tendency of poor women and women of color to nurse at lower rates than other women—without attention to the fact that, because of structural racism, classism, and sexism, many of these women work in jobs more prohibitive of pumping and expressing milk or attention to the fact that these women often live and work in places rendering their breastmilk more subject to environmental contamination. While the embracing of the "modern" and the "man-made" engendered the increasing reliance on formula from the 1930s to the early 1960s, the prominence of technology today plays into the attitude that science can fix whatever environmental problems we have created, along with the belief that environmental pollution represents a small price to pay for these "advances." The environmental contamination of breastmilk reminds us of the folly of our utilitarian approach to the "natural world" that sees diverse ecosystems as "resources" to be extracted and polluted, and which rarely questions our narrowness of vision.

Jared Diamond's recently published book *Collapse: How Societies Choose to Fail or Succeed*, about cultures that, despite great intellectual, cultural, and material achievement, fail and dissolve into the abyss of time offers some lessons. Diamond's book describes great civilizations that at one time appeared powerful and somehow ahead of their time—the Mayas, the Easter Island cultures, for example. But one quality that characterized these "collapsed" civilizations is a lack of foresight, and frequently a failure to recognize faulty environmental policy. These cultures, despite their gracefulness and vision, engineering feats, intellectual astuteness, and political and social dominance, failed to persevere and all went through periods of rapid decline. The warning to us, Diamond suggests, is that in our pursuit of wealth and material well-being, we are making choices that endanger our health, the health of our civilization, and the health of our planet. The failure to honor breastmilk, the choice to pollute it along with the other waters that inhabit our world, stands as a portent. It reminds us that we have a choice; we can continue in our blindness or we can see the consequences of our actions and move as a collective to do something to shift the balance.

ASSESSING COVERAGE: TWO WIDELY DIVERGENT APPROACHES

In conclusion, I want to focus on two recent pieces of attention to the threat to breastmilk and suggest some ways that we can move toward greater responsibility on this issue. In narrating this story, I have noted that recently several media sources have begun to bring much needed attention to breastmilk by starting to cover the reality of its toxicity and the contamination that gets imparted through placental fluid. Some of these articles have done a good job of starting to get the word out that breastmilk matters greatly, while also seeking to enlighten the public about the sensitive and complex problem of environmental toxicity. Others have highlighted contamination issues, without being very responsible in their handling of them. As I have suggested here, much of the controversy around the issue involves the fear on the part of advocates that any negative coverage will bring mothers to desist breastfeeding their babies.

I want to discuss a July 1, 2003, article appearing on MSNBCNews.com. Noting that concern has recently surfaced that pollutants get passed to fetuses and infants through the placenta and breastmilk, the article begins: "The government should teach women and girls to eat less of the fats found in meat, poultry, fatty fish and whole milk years before they become pregnant" ("Women Warned"). It points to an Institute of Medicine (IOM) report in which scientists working for the agency state that "perhaps the *most direct* way for an individual or a population to reduce dietary intake of DLCs is to reduce their consumption of dietary fat, especially from animal sources that are known to contain higher levels of these compounds" (my italics; 2). Because the MSNBC article offers findings from the report issued by the Institute of Medicine, part of the National Academy of Sciences, any blame over the handling lies there as much as with MSNBC's coverage. But, on the other hand, the reporting could have taken the findings one step further and offered commentary on the limitations of the report issued by the Institute.

I find it frankly baffling that a governmental institute could suggest that the primary way we should address this issue would be for women and girls to reduce their fat intake. Placing the burden on women and girls to address the complex issue of breastmilk toxicity and contamination that gets passed through placental fluid, suggesting that the answer to such problems lies in individuals regulating what they eat, and suggesting that the issue has little to do with the rest of the population—or with actions outside of individual eating choices—the article fails to address the larger problems and solutions at stake.[1] It fails to situate the findings, to offer some context about the importance of breastfeeding, or to note how scientists agree that, even when contaminated, breastmilk is still by far the better choice. And it fails to offer context about toxicity imparted through placental fluid and breastmilk and how such exposures occur. It does not address variations in exposure due to geography, occupation and age- and race-based exposures. And it suggests that the only behavioral shift that would be applicable would be for people to lower their intake of animal fats. Most problematic are its message that this is an issue only for women and girls, not for the population as a whole and its suggestion that the most appropriate response involves individual behavior modification, rather than individuals coming together on local, state, national, and international levels to push for changes in governmental policies, regulation of those policies, and system-wide changes that would benefit all of us. No wonder feminists, environmentalists, and other concerned citizens become outraged.

Some of the recent coverage of flame retardant exposures mentioned in the introduction of this book has done a good job of explaining that breastmilk toxicity is just one mechanism for assessing overall body burdens, and that just because breastmilk shows certain high levels does not mean that breastmilk should be seen as solely at stake. This is a step forward. But, at the same time, much of this coverage does not devote enough attention to the benefits of breastmilk or the systematic issues at stake. It is noteworthy that MSNBC, earlier, did offer the kind of responsible coverage that is needed. That article, "Concerns Raised on Breastmilk" begins: "the best food for a baby, physically and psychologically, is mother's milk, doctors agree" (Lyman). And it features a story of a mother, who also happens to be "a successful physician," sitting in a dark conference room watching "slide after slide flash . . . by on a huge screen . . . exposing a stark-looking roster of chemicals that read[s] like the manifest of a Superfund site." Balancing a focus on the benefits of breastmilk with a stern warning that if we do not act soon, the tides could turn, the article ends with a quotation from the mother, reminding us that this issue is about the right to information as well as the right for individuals and communities to have control over what goes into our bodies. It ends with the mother, who had eaten large amounts of Lake Michigan fish, proclaiming: "I would like to have known then what I know now. Everyone is entitled to this information yet this debate . . . is a debate taking place behind closed doors when it should be out and discussed among the lay public and among its doctors."

In the first article, the failure to even mention that a solution, if not the most *direct* solution, would be for us to put an end to the production and releasing of such chemicals into the environment becomes even more palpable when we confront the fact that those delivering the message represent a key governmental agency in charge of human health. It reminds us of how as a nation we fail adequately to address issues, in part because of the fragmentation of agencies, in part because of governmental ties to chemical and other industries and governmental unwillingness to take a stand against corporate interests. Another problem with the solution presented here involves the suggestion that the answer is for children to limit their fat intake. While we have all heard about the increasing tendency for children in the United States to be overweight, the fact remains that children do need a certain amount of fat in their diets to promote brain development and the development of other organs.

If we recall, some of the coverage of flame retardants do a much better job of making the case that toxic breastmilk is about much larger issues than the MSNBC coverage suggests. Many explain that scientists use breastmilk for biomonitoring because of convenience, and remind us that contamination levels for breastmilk indicate total body burdens, and that this is not just about exposures to breastfeeding women, but also about exposures to entire populations. On the other hand, indicating that environmental justice activism has begun to pay off, the MSNBC coverage of the IOM Report does offer a departure from earlier approaches in that it does address, if briefly, that Native American people face greater risks. Noting that "some American Indian tribes and indigenous groups in Canada are also at risk because they frequently eat fish and wild game," the panel at least mentions these most exposed groups. But it makes no mention of Native American or other sensitive populations in the United States, and it focuses only on diet, not on the larger question of why certain groups face greater exposures, or how blatant racism, sexism, ageism, and colonizing gestures all play a role. We need the kinds of cultural analysis offered by feminist and environmental justice groups to inform our cultural discourse. In suggesting that diet is the culprit, without contextualizing the importance of fishing and hunting to Native American and other sensitive groups, the article fails to broach just how disruptive it is to ask these groups to give up eating animal fat. It refuses to acknowledge the huge loss, culturally, for such communities to shift from animal diets, a shift that constitutes a burden on an entire way of life. Such an oversight becomes representative of racism and classism that continues to be felt.

THE CENTER FOR CHILDREN'S HEALTH AND THE ENVIRONMENT—BASICALLY APPROPRIATE, BOLD ATTENTION

Perhaps the most visible attention recently to breastmilk contamination appeared in a series of advertisements in the *New York Times* beginning June 5, 2002, and ending August 15, 2002.[2] While all the ads either refer specifically to environmental pollutants in breastmilk or point to more general exposures that can occur through breastmilk, the third in the series, titled "Our Most Precious Natural Resources Are Being Threatened. Why?," begins, in bold print, "Toxic chemicals are being passed on to infants in breastmilk."[3] The ads are notable in that they cover a full page—significant space—and include a large 8 x 10 photograph and bold type that catches our attention. Each ad also features the word "Why?" in bold print, inviting us to do something we are not often asked to—use our critical consciousness to question how things are, and ask ourselves whether we could do something individually and collectively to change the way things are.

Opening with "We've never created a product with the effectiveness of breast milk. Breast milk is a unique source of nourishment and protection against disease," the June 14 ad states "As pediatricians and scientists, we are convinced that breast milk is still the best choice for mother and child." The copy is careful to situate its findings, and to state from the outset that breastmilk offers much that cannot be replicated by formula. On the other hand, it does focus on risks. Stating that "we see disturbing evidence that in the future, breast milk may not be as effective as it once was in guarding against disease," it warns "Unless classes of chemicals that accumulate in breastmilk are phased out, we believe the health risks to our children could increase." With a tone of frankness—"We don't know what the minimum safe levels of exposure are. It may be that no exposure is safe"—the advertisements, while maintaining "we are convince that breast milk is still the best choice," do focus on "disturbing evidence." For example, the Center for Children's Health and the Environment, through the Mount Sinai School of Medicine, notes "several studies in the Netherlands show that as levels of PCBs in breast milk increased, infants had more immune impairment, evidence that toxic pollutants in breast milk can negate the milk's immunological benefits."

The ad is important in its boldness and visibility, in the way it speaks about breastmilk at great length, and in its frankness in pointing to the dangers associated with environmentally contaminated breastmilk. The promoters of the ads, scientists and doctors, add credibility to the issue in that they clearly demonstrate their expertise; each ad ends with "A summary of the supporting scientific evidence, and a list of scientific endorsers, can be found at www.chilenvironment.org." In the vein of Rachel Carson, the information is presented so as to convince the doubters among us, with detailed explanations, and a wealth of summaries of studies backing up the claims being made. For example, pointing out that while DDT has been banned, "today's breastmilk still contains toxic remnants . . . passed from grandmother to mother to child." And while noting that "a breastfed infant can absorb in one year thirty to ninety percent of the maximum recommended lifetime dose of dioxin, a chemical known to be both hormonally active and carcinogenic," it points to specific landmark studies to make its point. Like Carson, the tone is both even-handed and persuasive. While offering much that invites readers to stop and think about the implications, it also notes: "There is some good news as well." It then describes a Swedish study that demonstrated that as policy changes severely limited exposure to suspect chemicals there, the levels of the chemicals in breastmilk fell dramatically. In offering a link to the organization's website, the ads allow readers to become fully engaged in the effort to bring about change.

All the ads include a focusing headline—"Toxic chemicals can cause learning disabilities," "Toxic chemicals appear linked to rising rates of some cancers," "Pesticides could become the ultimate male contraceptive," and "Multiple exposures pose unknown risks"—a section called "What We Know," and a section discussing "What We Can Do." They thus speak to some pressing issues that many of us have heard of, but ones about which we may have little specific information. As such, they seek to draw a broad range of readers into their effort. I see much to commend about the ads, and believe that more attention of this kind must be brought to the issue, preferably featuring the bold visibility chosen by the ads' designers. Much, much more needs to be said to address the full scope of environmental pollutants and their effects. The ads repeadedly suggest that this is a children's health issue—a great start—but they do not address the fact that this is also very much an environmental justice issue as well as a women's issue. To do so would broaden the scope of the analysis. Of the seven photographs offered in the series of ads, only one includes a person of color, and this is ad #5, titled "Medicines are the only chemicals that have to be proven safe. Why?" It featured a picture, apparently, of an Asian or Asian American doctor. Given research suggesting that environmental pollution more pointedly affects people of color, Native Americans in particular, poor people, and people working with occupational exposures, it would make sense for the ads to feature more of these groups. The logo used by the groups responsible for the ad campaign—the Mount Sinai School of Medicine/Center for Children's Health and the Environment—on all of its web links does feature children of various racial and ethnic backgrounds, and a talk I attended sponsored by Nsedu Witherspoon, a spokesperson for the Children's Environmental Health Network, which is linked to the Center for Children's Health and the Environment, examined environmental racism, thus suggesting that the group is very much aware of the issue as an EJ one. Whoever made the decision about the *New York Times* ads may not have fully thought through the details of the images and implications at stake. As so many theorists working on issues of racism, classism, colonialism, and other questions of privilege have pointed out, any time we fail to address outright the presence of these problems, even if it is just an oversight, we have made a step toward forgetting our own privilege, and have become, to some degree, culpable (Russo, Smith "Racism").

On the other hand, the ads do a praiseworthy job of asserting that this is a systemic issue, and of urging us to take action. The ads make specific suggestions, such as reminding us to limit our exposure to pesticides, to dry cleaning that has not first been aired out, to fish known to be contaminated, and to paints, solvents, and cleaning products containing toxic and volatile chemicals. It is important that we bear in mind that exposure to pesticides takes many forms. A neighbor of mine buys organic fruits and vegetables, but not dairy products; she also often leaves a bottle of Round-up and a box of Miracle Gro on her kitchen counter. As I have reiterated here, scientists and activists have reminded us that we should not ignore urban and suburban, nonfarming use of pesticides and the risks they pose. So when we think about cutting our exposures to harmful chemicals, it is important that we prioritize both cutting out pesticides and chemicals used for home and pet pest control as well as buying organic foods—particularly organic dairy, fish, and meat products (because these sit higher on the food chain), then vegetables, grains, and processed foods. Paying attention to other products, such as those shown to contain PBCEs, can make a difference. Again, racism, classism, and colonization enter the picture when we recognize, as we are poised between the organic milk costing $4.25 and the conventionally produced, at $1.69 a gallon, that being able to purchase organic is a privilege that results from our society's system of economic disparities. And we should also recognize that in purchasing organic, we are impacting, not only our health and our children's health, but the health of farmworkers, workers in the chemical industry, and those living downstream from both. A good rule of thumb is to educate oneself on which fish offer better choices, to buy organic when possible, and to shop locally and at farmer's markets since such foods need fewer chemicals and preservatives because they require less transportation.[4] It is important to note that the Center for Children's Health and the Environment ads do not stop with addressing our eating and purchasing habits. They boldly assert: "We must do more." Unlike the MSNBC article, they address these as societal problems, not as ills to be bandaided by individual eating choices. After asserting that "we must phase out chemicals that pose a risk to our health, especially to our children's health, beginning with the toxic chemicals that have been detected in breast milk," they further urge: "We should demand that new chemicals undergo the same rigorous testing as medicines before [they are] allowed on the market."

Educating readers who may not be aware of current government policies, they note that while "we don't allow food or drugs to be sold before being shown to be safe," we do allow thousands of chemicals on the market, chemicals that affect human biology, endocrinology, and reproductive health, chemicals that have never been tested in isolation, much less in combinations that are suspected to contribute to high cancer rates, lower fertility rates, neurological impairment, and other health issues. Flatly asserting that "we do not have a system that does that now," they advocate "we must institute a system of regulation that tests new synthetic chemical and proves them safe before they are allowed to be sold" (Ad #1—June 5; Ad #7—August 15). Asking, "Isn't that a system you thought we already had?" they imply directly to the trust issue. They imply a long-enduring fact about environmental awareness—that many Americans, trusting that the government would not permit anything that was really harmful to be released into the environment, become involved in environmental activism only to a certain, limited degree.

When it comes down to it, we do not want to believe that we are eating, drinking, and breathing toxic chemicals, and we cannot see evidence that we are. Even cancer can be attributed to other factors, to behavior, "lifestyle," or genetics. So many of us go on hearing environmentalists' messages as important—up to a point. But then, a line is drawn. At some point, we harbor the belief that environmentalists are a bit over the top, a bit crazy, a bit overblown. At some point, we tire of messages that seem to suggest that everything is toxic; we harbor a nostalgia for the good old days when we did not question and doubt. And we throw up our hands, believing that, in the end, we are safer than previous generations. These ads do much to shake us out of our complacency, to force us to start questioning what is going on, and to see that we can make a difference.[5] They spark us to take action.[6] The breastfeeding ad reminds us that toxic breastmilk illuminates our position at an impasse when it asserts: "There can be no more important public health mission than ensuring the safety of mother's milk." It invites us to start, today, to work to change the system that we have sanctioned though our silence.

NOTES

1. The specific health effects noted by the advertisements include: lowered risk of ear infections, upper respiratory tract infections, and certain forms of cancer. The ads also point to the recent (May 2004) study by the National Institutes of Health.

2. See Appendix A for comparative studies of rates of breastfeeding.

3. One mother I spoke with explained that she had struggled to continue nursing her child, but was unable to do so after her baby reached seven weeks. After speaking with her, I believe that the advertisements would help her and women like her to feel more comfortable if they had emphasized that any amount of breastmilk can be helpful (Peterson). According to the Ad Council Newsletter, babies not breastfed have a 30 percent increased risk of developing leukemia and up to a 40 percent increased risk of developing diabetes (Peterson).

4. California's bill requires that participants "receive consultation, health care referrals, follow-up counseling, educational activities and materials" (Patton).

5. An eighteen-year study conducted in New Zealand found that breastfeeding is a "significant predictor of later cognitive and educational outcome, even after adjusting for differences in socioeconomic and health status" (Steingraber "Having" 240).

6. A 2002 study appearing in *JAMA* found that the "duration of breastfeeding was associated with significantly higher scores on the Verbal, Performance, and Full Scale WAIS IQs" (Wechsler Adult Intelligence Scale) (Mortensen et al. 365–46). Indeed, "With regression adjustment for potential confounding factors, the mean Full Scale WAIS IQs were 99.4, 101.7, 102.3, 106.0, and 104.0 for breastfeeding durations of less than one month, 2 to 3 months, 4 to 6 months, 7 to 9 months, and more than 9 months, respectively (2365). The study concludes that "significant positive association between duration of breastfeeding and intelligence was observed in 2 independent samples of young adults, assessed with 2 different intelligence tests."

7. According to "Business Backs Breastfeeding," companies that adopt breastfeeding support programs note cost savings of up to $3 for every $1 invested in breastfeeding support, less illness among breastfed children of employees, reduced

absenteeism to care for ill children, employee retention, and family-friendly image in the community, among other benefits (9).

8. The recent biomonitoring studies have already had an effect on policy. For example, Governor Gray Davis signed legislation in August 2003 to make California the first state in the nation to institute a PBDE ban; other states have begun to phase out the chemicals since then (Pohl).

9. Swedish bans have brought about changes in body burden of PBDEs in Swedish women; for example, a 1998 study showed that levels in Swedish breastmilk doubled every two to five years, but later studies conducted after the ban showed that levels quickly began to decline. The environmental working group's study found a median level of 58 parts per billion in US women, compared with a median level of 2 parts per billion in a Swedish study from 2002 (Shipley). Other studies have followed (Schmidt).

10. It is important to note that in many cases alternatives to PBDEs have been found. For example, the EPA has given preliminary approval to a Penta (one type of PBDE) substitute called Firemaster 550, which the agency says does not build up in the environment (Shipley). The EPA has announced that it would discuss a voluntary phase-out with other manufacturers. Great Lakes Chemical Corp., for example, said in 2003 that it would replace two widely used brominated fire retardants, penta and octa, with a product deemed safer. A growing list of companies—IKEA, Motorola, Ericsson, Intel—have redesigned their products to eliminate the need to add PBDEs (Schmidt).

NOTES TO CHAPTER 1—HISTORICAL VIGNETTES

1. I borrow this term "corporatization" from Stephen Hawking.

2. Palmer comments about of the idea that semen curdles milk: "This concept of semen entering the milk is not as ridiculous as it may seem, for as stimulation of the breasts may be followed by sensations in the pelvis and genitals the connection between the two areas of the body and the two white substances is an obvious one" (152). (See Valerie Fildes, *Breasts, Bottles, and Babies*, 104, for an opposing view.)

3. Antonia Fraser states "Sexual relations between the wet-nurse and her husband (or anyone else) were also [in addition to those between the husband and wife of higher social ranks] considered dangerous but it is highly doubtful that such prohibitions actually brought about celibacy" (77).

4. Janet Golden, looking at the business records from 1683 to 1703 of John Pynchon of Springfield, Massachusetts, notes that he retained two nurses and over thirty domestics and that the wet nurses Good Wife Taylor and Hannah Excell earned twelve shillings a month—more than Pynchon paid his other female servants (28). Gabrielle Palmer, studying seventeenth-century England, notes that "pay was quite good" and that "wet nursing was like the catering trade: your status was derived according to whom you feed" (160).

5. It should be noted that, in some cases, what was called the transmission was an indirect "transmission"; for example, in the case of syphilis, the disease was not directly transmitted through the breastmilk; rather, it was the result of poor hygiene or contact with lesions or chancres (Palmer 169).

6. Today we have six human milk banks in the United States and one in Canada.

7. For a slightly different reading of this, see Palmer 144–45.

8. Enfamil LIPIL with iron claims that "Only Enfamil has LIPIL, a unique blend of DHA and ARA, important nutrients found in breast milk that support brain and eye development. And Enfamil LIPIL with iron has levels of DHA and ARA similar to worldwide breastmilk" (Enfamil LIPIL). Nestlé Good Start claims: "All DHA &ARA formulas are designed to support baby's brain and eye development with nutrients found in breastmilk, baby's ideal food. But Good Start Supreme DHA & ARA is the only one made with comfort proteins that are broken down into smaller pieces to be easy on your baby's tummy" (Nestlé).

9. The three major competitors are Ross, with its Similac brand, Mead Johnson with its Enfamil brand (which make up 90% of U.S. formula sales), and Wyeth with SMA.

10. See Appendix A for estimated rates of breastfeeding worldwide.

11. For updates on the Nestlé boycott and issues surrounding the direct marketing of infant formula, see www.babymilkaction.org.

12. This approach is from *Blueprints Obstetrics and Gynecology*, a "standard" textbook used by medical students, residents, and other healthcare practitioners. The passage continues: "Increased transmission can be seen with higher viral burden or advanced disease in the mother, rupture of the membranes and events during labor and delivery that increase neonatal exposure to maternal blood" (Canghay et al. 81).

13. In 1985, when researchers first discovered HIV in mother's milk, US Public Health Service officials issued recommendations applicable only to the United States that infected mothers not breastfeed (Altman "AIDS").

14. The JAMA article is "HIV Transmission Through Breastfeeding" by Paolo G. Miotti et al. The three year study is "Effect of Breastfeeding and Formula Feeding on Transmission of HIV-1: A Randomized Clinical Trial." For research about HIV transmission through breastmilk, see also Bassett; Choto; Dabis, Leroy, Castetbon et al.; Fowler, Bertolli, and Nieburg; Frieden; Goldsmith; Guay and Ruff; John, Richardson, Nduati et al.; Nduati et al.; Taha et al.; and Van de Perre.

15. The editorial mentioned here is "When Is Breastfeeding Not Best?: The Dilemma Facing HIV-Infected Women in Resource-Poor Settings."

16. Indeed, a relatively recent (2002) article notes that while "free formula may appear to be a blessing . . . it is very likely to increase morbidity and mortality from other infectious diseases, thus decreasing overall child survival" (quoted in Dobson 1474). The authors note that "even in areas of high HIV prevalence, [they] believe it is more appropriate to promote exclusive breastfeeding as public health policy." The article quoted appears in *Health Policy and Planning* (2002): 154–60.

17. A study entitled "Breastmilk Transmission of HIV-1: Laboratory and Clinical Studies" notes that "breastmilk contains immunoactive cells, antiinfectious substances, immune globulins [and] cytokines" (Van de Perre 122).

18. The *Wall Street Journal* article referred to here was printed on 5 December, 2000.

19. Baby Milk Action's update twenty-nine states that "it is a common misconception that all women diagnosed as being infected will pass the virus to their infants." It points out that "of 100 women in a community with 20% HIV prevalence among women at delivery," twenty of the women will be infected with HIV, with thirteen of these being likely to pass the virus to their infants, and four likely to pass

the virus during pregnancy or at birth, and only three likely to pass the virus through breastfeeding" (Update 29).

20. One article discusses how HIV-positive Kathleen Tyson, mother of Felix, found herself on trial after a court took legal custody of her infant when she declined to accept the court's order that she desist breastfeeding (Kent). Kent notes that "several state governments have acted to override parents' decisions and insisted that the infants of HIV-positive mothers must be subjected to treatments with various forms of antiretroviral therapy," and "some have refused to allow the newborn infants to be breastfed." Kent argues that if these mothers are to be stopped from breastfeeding, they should be informed of the full range of options, including expressing and heating one's own milk (which has been shown to reduce transmission) using banked milk, or using a wet nurse. It should be noted that the HMBANA advises against co-nursing, as the transmission of conditions can take place in such situations.

21. Linda Greenhouse, reporting on the case in the *New York Times*, notes that "the question was whether the drug tests, conducted without warrant and without suspicion of individual wrongdoing, were constitutional searches." Pointing out that justices' responses "appeared to depend on whether [they] saw the case through the prism of criminal law or public health," Greenhouse notes that "it was clear from their questions that at least some of the justices saw the case in a larger framework of the debate over what steps society may take in response to risky behavior by pregnant women." Justice Ruth Bader Ginsburg's comment that "I don't see a 'protective purpose' because many of the women were arrested after their babies were born" is the closest anyone came to considering how the case also has bearing on whether such women would be able to nurse their babies or not.

NOTES TO CHAPTER 2—TOXIC DISCLOSURE

1. The *New York Times* magazine featured an article on the subject on January 9, 2004 by Florence Williams.

2. These figures were obtained by conducting a Lexus-Nexus search. My search strategy involved questing for the following keywords anywhere in the article: "breastfeeding" or "breast feeding" or "breastmilk" or "breast milk" or "nursing" or "infant feeding" or "human milk" and "pollution" or "toxins" or "contamination" or "DDT" or "POPs" or "carcinogens" or "dioxins" or "mercury" or "PCBs" or "environment."

3. The Toxic Substances Control Act was enacted in 1976; Superfund in 1980.

4. See Appendix A for more on rates.

5. For example, the northeastern region ran seven articles on the issue, between December 7, 1986, and March 9, 1986. Between December 26 and March 8, 1986, papers in the southeast region ran twenty-four articles on various aspects of the debacle, including articles on the dumping of contaminated milk, the testing of breastmilk, the funding of these tests, and the prosecution of those believed responsible for the poisoning.

6. Often these problems are caused by a yeast infection that can simply be treated by giving both mother and baby easily administered prescriptions.

7. The environmental contamination of infant formula enters the picture in 1999, in Europe, when the European Commission passes a Directive setting limits on pesticides in baby milks (113).

8. The *Times* notes that the writer of the piece, Lewis Regenstein, vice president of the Fund for Animals, is author of "America the Poisoned: How Deadly Chemicals Are Destroying Our Environment, Our Wildlife—and Ourselves."

9. On April 16, 1989, the *New York Times* published another article featuring *Silent Spring*. The piece, "Passing on the Legacy of Nature," looks at how a father taught his daughter, and how she, reading *Silent Spring*, learned of breastmilk contamination, among other effects of our disregard for the environment.

10. The *Times* article points out that a study conducted in 1987 at Laval University in SteFoy, Quebec, found that PCB concentrations were on average 3.59 parts per million in Eskimo women compared with .76 parts per million for the rest of the samples from an industrial suburb of Quebec City, with some Eskimo women showing concentrations of 14.7 parts per million. While the article does note that the PCBs in the area "enter the environment from waste poured into rivers and coastal waters could come from as far away as the Soviet Union," it does not discuss the process, or the effects of bioaccumulation. And in noting that "we Southerners eat a lot less fish than the Inuit," whose diet includes "large amounts of fish and animal fat," it suggests that we southerners are immune to such effects.

11. Harrison, noting that "the story gave less prominence to breast milk as a source of exposure," also notes that "in other stories, the *New York Times* focused on dioxin in cow's milk, bleached coffee filters, and contaminated fish, and did not even mention breast milk" (43). She cites as evidence Philip Shabecoff, who states: "Government says dioxin from paper mills poses no danger."

12. And even if one does not consume fish, that the fish in one's immediate environment is subjected to environmental pollutants attests to the fact that one's environment as a whole is not being treated in a way that would lead to optimal health and well-being.

NOTES TO CHAPTER 3—THE OMISSIONS OF BREASTFEEDING AND BREASTMILK CONTAMINATION AS SIGNIFICANT FEMINIST ISSUES

1. See No. 122, January/February, 2004.

2. This search was conducted using Infotrac's Expanded Academic Index ASAP; the publication listing what the NWSA considers to be key "Women's Studies Journals" is entitled "Defining Women's Studies Scholarship: A Statement of the National Women's Studies Association Task Force on Faculty Roles and Rewards" (Pryse). I did not specify dates for the Infotrac search; entries begin in 1980, and the list was last updated June 2002.

3. See http://www.now.org.

4. The report by the Family, Gender and Tenure Project at the University of Virginia, called "Parental Leave in Academia," was based on a national study of 168 institutions and was completed in 2001 by Charmaine Yoest, a graduate student, and Steven E. Rhoads, a UVA professor of politics. The study was financed by the Alfred P. Sloan Foundation and the Bankard Fund for Political Economy at the University

of Virginia. It is available at http://faculty.virginia.edu/familyandtenure/ institutional%20report.pdf.

5. My husband, who is beginning his medical residency as I complete this chapter, tells me that his medical training involved only a "couple of minutes" of time devoted to the benefits of breastmilk. The training also involved attention to the recommendation that HIV-positive mothers not breastfeed, given the possibility of transmitting the virus to their infants.

6. An article published in *Pediatrics* in 2001 points out that "when compliance with the Ten Steps [part of the Baby Friendly Initiative] is achieved, the results are dramatic"; Evergreen Hospital in Kirkland, Washington, the first baby friendly hospital in the United States, has an initiation rate of more than 90 percent ("Physicians" 588).

7. Linda Layne makes a similar claim about "pregnancy loss," arguing that because feminists have not embraced the problem of miscarriage like they have other issues such as the medicalization of childbirth, women who experience miscarriage do not have the benefit of a feminist framework to inform their experiences of loss ("Breaking" 296).

8. Bernice Hausman does discuss Van Esterick in *Mother's Milk: Breastfeeding Controversies in American Culture.*

9. Several sources offer advice to adoptive mothers on developing a milk flow. For example, see Horowitz.

10. For rates, see Appendix A. While the number of mothers participating in the workforce has gone up significantly in recent years, a *New York Times* article from 2002 points out that "[a]fter a quarter-century in which women with young children poured into the workplace, the percentage of women in the labor force who had babies younger than 1 year old declined last year" (Lewin).

11. Breastfeeding initiation rates listed by one source are: mothers who are college graduates (80.1%), mothers over thirty-five years old (74.7%), Anglo-American mothers (70.2%), Hispanic mothers (69.4%), mothers employed full-time (65.6%), teenaged mothers (54.4%), mothers with "less than a high school education" (52.2%), African American mothers (50%) (Platypus Media).

12. The same survey compared rates for other racial and ethnic groups, looking at percentages of mothers who breastfed even for a brief period: 45 percent of African-American mothers, 59 percent of Native American mothers, 66 percent of Latina mothers, 68 percent of "white" mothers; and 54 percent of low-income Asian and Pacific Islander mothers (Appea 1).

13. It is telling that in her discussion of the few feminists who have addressed breastmilk, Hausman does not mention Sandra Steingraber's work on breastmilk, or other considerations of breastmilk toxicity; her characterization of feminists writing on the issue along disciplinary lines, which does not include ecologists, and her omission of ecological issues in her overview of feminist attention to the issue suggest why. I focus on Hausman because she has published most recently and her work can be considered in line with other feminist attention.

14. The study found that "duration of breastfeeding was associated with significantly higher scores on the Verbal, Performance, and Full Scale WAIS IQs." The adjusted mean Full Scale WAIS IQs were 99.4 for breastfeeding duration of less than one month, 101.7 for duration of 2 to 3 months, 102.3 for duration of 4 to 6 months, 106.0 for duration of 7 to 9 months, and 104.0 for duration of more than 9 months.

15. Proposing two models of infant feeding, Van Esterick notes that the "infant formula model" involves a "nonrenewable resource," an "inert substance," calcu-

lated on the "scarcity principle," in which dependency and consumerism get enforced, while the "breastfeeding model" involves a "renewable resource," a "living product," calculated on the "satiation principle," in which the infant is empowered, active, and in control (*Beyond* 201).

16. The interview appeared in the March, April 1998 edition; letters from readers expressing concern appeared in later editions (Snell).

17. It should be noted, here, that eleven of the sixteen respondents either rejected the label "environmentalist" outright or described themselves as an "ecologist" or as "new," or "third world," or "indigenous" environmentalists (Prindeville and Bretting 50).

18. The policy statement argues that "human milk is uniquely superior in infant feeding," and that "all substitute feeding options differ markedly from it" (Gartner 19). It notes that benefits extend not only to individual infants, mothers, and families, but also to employers in terms of economic savings from reduced absenteeism, and to the nation from reduced health costs.

19. Later editions of *Our Bodies, Ourselves* mirror the later statements by NOW in that they advocate breastfeeding and seem less concerned about women being coerced into nursing and more focused on convincing mothers to take the breast-versus bottle question seriously, offering support, for example, for women wanting to breastfeed and return to work, and suggesting that La Leche League offers information on renting breast pumps (Boston Women's 507–10). The reason given in the earlier edition for why it is dangerous to prompt women to breastfeed is that such arguments "do not take into account the possible disastrous effects of a mother feeding in a way she does not wish or cannot do."

NOTES TO CHAPTER 4—ENVIRONMENTAL RACISM, ENVIRONMENTAL JUSTICE, AND BREASTMILK CONTAMINATION

1. It is the fat, not the liquid, that becomes the repository for harmful chemicals.

2. In addition, "Children have greater exposures to environmental toxins than do adults," since "pound for pound of body weight, children drink more water, eat more food, and breathe more air than do adults" (Center 1). For example, children ages one to five years consume three to four times more food per pound than do the average adult in the United States. Children thus have substantially higher exposures pound for pound than adults. And this "has been demonstrated clearly in the case of children's exposures to pesticides in the diet" (1).

3. Schell and Tarbell, concurring with perspectives such as Uma Narayan's, Jacqui Alexander's, and Chandra Talpade Mohante's who warn against colonizing gestures on the part of researchers, articulate the importance of being aware of each "particular community's cultural identity" (834). They note "Each community is unique and special in its own way. Even populations of the same ethnic group may differ. For example, there are many different native communities across the Americas and Canada, but each is unique. One cannot assume that all native communities across the United States are the same. . . . This is true for other minority communities" (834).

4. Ms. Cook's name, pronounced "Gudgi," gets spelled both as Gudgi and Katsi.

5. According to Dr. Sears, considered by many to be an expert on child care, "Breastfeeding families save $600 to $2,000 a year, and often much more in medical

bills since baby stays healthier and employed breastfeeding mothers miss less work" (Ask Dr. Sears.com). He notes that hypoallergenic formulas cost up to $2,000 a year.

6. According to the "Report on the Mt. Olive Pickle Boycott," these workers have been excluded from much federal and state legislation, including the National Labor Relations Act (NLRA) of 1935 and the Fair Labor Standards Act (FLSA) of 1938, which means that farmworkers face some of the worst conditions of all U.S. workers. While the FLSA was amended in 1966 to cover farmworkers, it excluded small farms from minimum wage provisions, while other pieces of legislation excluded farmworkers from pension plans, unemployment insurance, and workers' compensation ("Report on Mt. Olive" 2).

7. Right -to- know laws do not govern toxicants used in agriculture, except in California and New York, meaning that pesticides are not included in figures for toxic releases.

8. The study also "analyzed new data from a 1998 study that reported a 33 percent increase in the risk of non-chromosomal birth defects, such as cleft palates."

NOTES TO CONCLUSION

1. As such, it mirrors much coverage of cancer incidences that if they broach the question of probable causes, fail to mention that environmental factors might very well be at stake, but instead focus on "lifestyle issues," including dietary changes that individuals might take in order to positively impact outcomes.

2. To see the advertisements and the Position Papers that provide scientific background and sources for them, visit the Center for Children's Health and the Environment Web Site http://www.childenvironment.org/position.htm. The titles of the advertisements are as follows: (1) "Johnny Can't Read, Sit Still, or Stop Hitting the Neighbor's Kid. Why?" (2) "More Kids Are Getting Brain Cancer. Why?" (3) "Our Most Precious Natural Resources Are Being Threatened. Why?" (4) "Pesticides Could Become the Ultimate Male Contraceptive. Why?" (5) "Medicines Are the Only Chemicals That Have To Be Proven Safe. Why?" (6) "Chemicals Combine in Our Bodies But Are Rarely Treated That Way. Why?"

3. This ad, the third in the series, appears on page A19.

4. On contamination through eating fish, see Appendix B.

5. It is important to note that one study, published in the *New England Journal of Medicine*, looked at the question "Does popular coverage of medical research in turn amplify the effects of that research on the scientific community" (Phillips, Kanter, Bednarczyk et al. 1180–83). Its findings, that "coverage of medical research in the popular press amplifies the transmission of medical information from the scientific literature to the research community," suggests the added importance of coverage such as the ads by Center for Children's Health and the Environment.

6. If one goes to the website offering more specifics on "What can I do?," several suggestions are offered, including one that "If our society is to achieve the goal of a safer environment for our children and our children's children, we must begin by recognizing and supporting children's environmental health initiatives on a local, state, and federal level" (www.childenvironment.ort/parent_text.htm). Focusing on the need for individuals to come together, it asserts: "This involves active community participation on the local level, recognition and promotion of state and federal legislation that supports children's environmental health initiatives and monetary support to those agencies and institutions that are workings towards these goals."

APPENDIX A

There is some disagreement about rates of breastfeeding in the United States and abroad. Discrepancies may involve whether studies are considering "exclusive" breastfeeding and the duration of breastfeeding being referred to. Some studies consider initiation rates—including whether a woman breastfeeds even for a few days. According to the World Health Organization, "exclusive" breastfeeding allows drops and medicinal syrups, while "predominant" means that the "prominent source of nourishment has been breastmilk."

Currently, between 70.1 percent and 53 percent of all mothers in the United States attempt breastfeeding—at least once (Platypus Media). But these numbers fall off significantly by six months, if not well before, especially if one is concerned with exclusive breastfeeding. An article appearing in *Pediatrics* in its May 2003 Supplement notes that "almost two thirds (65.1%) of children had ever been breastfed." Approximately 60 percent of infants are exclusively breastfed in the early postpartum period, and just 7.9 percent are exclusively breastfed by six months (3). At six and twelve months, 27.0 percent and 12.3 percent, respectively, were receiving some breast milk" (Ruowei et al.). Only 17 percent exclusively breastfed for six months (Wolf 2007; Ruowei et al.)

Abbott Laboratory's ongoing "Feeding Infants and Toddlers Study" (FITS) notes that 70.1 percent of mothers now initiate breastfeeding in the hospital and 33.2 percent are still breastfeeding when their babies reach six months. Again, these figures do not reflect exclusive breastfeeding. Fifty-three percent of lactating mothers introduce formula before their babies are a week old, 68 percent do so by 2 months, and 81 percent by 4 months (Wolf 2007; Ruowei et al.)

Sandra Steingraber notes that "today, the United States has one of the lowest breastfeeding rates in the world" and that "only half of all mothers attempt to initiate nursing at birth, and less than 20% continue at six months" (*Having* 245).

Breastfeeding initiation rates by groups in the United States are as follows:

Mothers who are college graduates	80.1%
Mothers over thirty-five years of age	74.7%
Anglo-American mothers	70.2%
Hispanic mothers	69.4%
Mothers employed full-time	65.6%
Teenage mothers	54.4%
Mothers with less than high school education	52.2%
African American mothers	50.0%

(Source: http://www.platypusmedia.com/adult_rates.html)

According to UNICEF, "The percentage of babies who are exclusively breastfed for the first four months of life varies from 90% to 1% depending on where the baby is born." For figures throughout the world, see Breastfeeding League, UNICEF.org.

APPENDIX B

Several sources have issued warnings about fish contaminated with PCBs, mercury, dioxin, pesticides, and other toxic chemicals and have noted that scientists link these contaminants to brain damage, hormone disruption, cancer, and other health problems. Many of these toxins accumulate in the body and can be passed on to the fetus during pregnancy or to the infant during breastfeeding. CHEC Health offers information and suggestions for how best to minimize exposure to toxins that accumulate in fish (Children, Health Environment Coalition). They also offer a "Safe Fish Chart" that breaks fish down into categories, such as "Frequent Consumption 2–3 times per week" (farmed rainbow trout, trap-caught shrimp, and farmed clams), "Once a Week" (farmed catfish, Pacific cod, haddock, hook-caught Mahi-Mahi, farmed oysters if not from the Gulf of Mexico, perch, Pacific Pollock, and farmed sea bass), "Once a Month or Less" (Mussels, and Mackerel), and those fish to "Avoid" (freshwater bass, bluefish, halibut, great lakes fish, tuna steaks and canned white albacore tuna). The site also offers lists of fish that have been overfished and should be avoided ("CHEC's Safe"). Beginning in 2004, CHEC offered Spanish language lists and information for Spanish speakers, an important move given the high percentage of Spanish language speakers relying on fish for dietary needs ("Spanish CHEC").

For more on fish advisories, see "Fish Advisories" on the U.S. Environmental Protection Agency Website http://www.epa.gov/ost/fish/. Also see the Clean Air Task Force's report "Mercury, Power Plants and the Fish We Eat," issued in June 2000.

REFERENCES

Acury, Thomas A. et al. "Pesticide Safety Among Farmworkers: Perceived Risk and Perceived Control as Factors Reflecting Environmental Justice." *Environmental Health Perspectives Supplements* 110.2 (April 2002): 233–241.

Ad Council. "Ad Council: Breastfeeding Awareness." http://www.adcouncil.org/campaigns/breastfeeding/.

Aesoph, Lauri M. "Breastmilk: The Perfect Food." *Naturopathic Medicine*. Healthworld Online. 1994. http://www.healthy.net/Library/Articles/Aesoph/BreastMilk.htm.

"African Breastfeeding Rates Threatened." Infactcanada. http://www.infactcanada.ca/african_breastfeeding_rates.htm.

Akwasasne Task Force on the Environment. "Superfund Clean Up of Akwasasne: Case Study in Environmental Justice." *International Journal of Contemporary Sociology* (October 1997).

Alexander, M. Jacqui, and Chandra Talpade Mohante, eds. "Introduction: Genealogies, Legacies, Movements." *Feminist Genealogies, Colonial Legacies, Democratic Futures*. New York: Routledge, 1997. xiii–xlii.

Alston, Dana. "Transforming a Movement." *Race Poverty & the Environment*. II. 3 & 4 (Fall 1991/Winter 1992): 27–29.

Altman, Lawrence K. "AIDS Brings a Shift on Breast-Feeding." *New York Times* 26 July 1998, late edition, final: Foreign Desk section 1.

———. "Doctor's World; Childhood Death; Respiratory Ailments Are Now No. 1 Cause." *New York Times* 8 April 1986, late city final ed.: Science Desk section 1.

American Academy of Pediatrics. "Breastfeeding and the Use of Human Milk." Policy Statement. *Pediatrics* 100, no. 6 (December 1997).

Anderson, J. W. et al. "Breast-Feeding and Cognitive Development: A Meta-Analysis." American Journal of Clinical Nutrition 70 (1999): 525–535.

"An Easy Guide to Breastfeeding for African-American Women." July 2001. http://www.4woman.gov/owh/pub/aabreastfeeding/important.htm.

Appea, Pamela. "Nursing vs. Bottles: Why Are African American Breastfeeding Rates so Low? http://www.africana.com/DailyAricles/index_20001206.htm.

Apple, Rima D., and Janet Golden, eds. *Mothers and Motherhood: Readings in American History*. Columbus: Ohio State UP, 1997.

Arbuckle, T. E., and L. E. Sever. "Pesticide Exposure and Fetal Death: A Review of the Epidemiological Literature." *Crit Rev Toxicol* 28 (1997): 229–270.

Arkansas Health Department. "CDC Offer Good News for Milk Drinkers." *Arkansas Democrat-Gazette* 11 April 1986.

Arnold, Lois W. "Where Does Donor Milk Banking Fit in Public Health Policy?" *Breastfeeding Abstracts* 19.3 (2002): 19–20.

Arora, S. et al. "Major Factors Influencing Breastfeeding Rates: Mother's Perception of Father's Attitude and Milk Supply." *Pediatrics* 106.5 (2000): E67–68.

Arquette, Mary. "First Environment Restoration Initiative." Environmental Justice Project Description-DERT. National Institute of Environmental Health Sciences: Division of Extramural Research and Training. 6 October 2003. http://www.niehs.nih.gov/translat/envjust/projects/arquette.html. (12/7/2003).

Arquette, Mary, Maxine Cole, Katsi Cook et. al. "Holistic Risk-Based Environmental Decision Making: A Native Perspective." *Environmental Health Perspectives Supplements* 110.2 (April 2002): 259–265.

Aseplin, A. L. Pesticide Industry Sales and Usage: 1994 and 1995 Market Estimates. Washington, DC: U.S. Environmental Protection Agency, 1987.

Ask Dr. Sears.com. "Bottle-feeding Index." www.askdrsears.com/html/.

Baby Milk Action. Update 25. http://www.babymilkacton.org/update25.html.

———. Update 29. http://www.babymilkaction.org/update 29.html.

Bailey, Ronald. *Ecoscam: The False Prophets of Ecological Apocalypse.* New York: St. Martin's, 1993.

Bassett, M. T. "Psychosocial and Community Perspectives on Alternatives to Breastfeeding." *Annals of the New York Academy of Sciences* 918 (November 200)L128–135.

Baumslag, Naomi, and Dia L Michels. *Milk, Money, and Madness: The Culture and Politics of Breastfeeding.* Westport, CT: Bergin and Garvey, 1995.

Beach, Wooster. *The American Practice Condensed, or, the Family Physician.* 14th ed. New York: James McAlister, 1848.

Beadle, J. *The Journal or Diary of a Thankful Christian.* London, 1656.

Birkland, T. "Focusing Events, Mobilization, and Agenda Setting." *Journal of Public Policy* 18 (1998): 53–74.

Blum, Linda. "Mothers, Babies, and Breastfeeding in Late Capitalists America: The Shifting Contexts of Feminist Theory." *Feminist Studies* 19.2 (Summer 1993): 291–311.

Bonner, Alice. "Improper Use of Baby Formula Not Only An Overseas Problem; Misuse of Infant Formula Not Confined to Overseas." *Washington Post* 19 June 1981, final ed.: MetroB1.

Boston Women's Health Collective. *Our Bodies, Ourselves for the New Century.* New York: Simon & Schuster, 1998.

Bottorff, J. L. and J. M. Morse. "Mother's Perceptions of Breastmilk." *Journal of Obstetric, Gynecologic, and Neonatal Nursing* 19.6 (1990): 518–527.

Breast Cancer Action. "BCA Responds to the September 23rd Senate Hearing on Breast Milk Monitoring." http://www.bcaction.org/Pages/LearnAboutUs/BreastMilk.html.

"Breastfeeding and Environmental Contamination: A Discussion Paper." Nutrition Section. New York: UNICEF, May 1997. 1–12.

"Breastfeeding and HIV." Media Release. La Leche League International. 4 July 2001. http://www.lalecheleague.org/Release/HIV.html.

"Breastfeeding Practices in Los Angeles County." L. A. Health: A Publication of the Los Angeles County Department of Health Services Public Health. February 2001.

"Breastfeeding Rates." 2001. http://www.platypusmedia.com/adult_rates.html. (20 April 2002).

"Breastfeeding Rates Are on the Rise: Healthy People 2000 Research Shows Increase." All About Breastfeeding—Topics and Questions. 1998–2000. http://www.breastfeeding.com/all_about/all_about_bf_rates.html.

"Breastmilk Attacks." Infact Canada. http://www.infactcanada.ca/BMAttacks.html.

"Breastmilk: The Perfect Renewable Resource." Spring 1997. http://www.infactcanada.ca/ren_res.htm. (29 May 2002).

"Breast Milk Tests Show Pesticide Level at Norm." Arkansas Democrat-Gazette 29 April 1986.

Bristol-Myers Squibb Company Home Page. http://www.bms.com/public/aidsba.html (July 1999).

Brown, Warren. "The Role of Infant-Formula Makers in Developing Nations Hit." Washington Post 24 May 1978, final ed.: A8.

———. "Scientists Warn of Contaminated Milk for Infants." Washington Post 9 June 1977, final ed.: Metro C11.

Brozan, Nadine. "Two U.S. Aides Resign Over Baby-Formula Vote." New York Times 21 May 1981, Late City Final Ed.: Section A, page 9.

Bryant, Bunyan, ed. Environmental Justice: Issues, Policies, and Solutions. Washington DC: Island Press, 1995.

Bullard, R. D., ed. Dumping in Dixie: Race, Class, and Environmental Equity. Boulder: Westview Press, 1990.

———. "Environmental Justice for All." Unequal Protection: Environmental Justice and Communities of Color. Ed. Robert D. Bullard. San Francisco: Sierra Club Books, 1994. 3–22.

Burg, J., and G. Gist. "Observations from the CDC: The Potential Impact on Women from Environmental Exposures." Journal of Women's Health 6.2 (1997):159–161.

Burros, Marian. "Ecology at Home." Washington Post 7 July 1977, final ed.: Style Food, E1.

Business Backs Breastfeeding: A Flexible Workplace Program for Breastfeeding Mothers. Ross Products Division, Abbott Laboratories. http://www.ross.com.

Carson, Rachel. Silent Spring. Boston: Houghton Mifflin, 1962.

Carter, Pam. Feminism, Breasts and Breast-Feeding. New York: St. Martin's, 1995.

Caughey, Aaron B., Tamara L. Callahan, Linda Heffner. Blueprints Obstetrics and Gynecology. NY: Blackwell Publishers, 2003.

Center for Children's Health and the Environment. "Children's Unique Vulnerability to Environmental Toxins." (June 1, 2002). http://www.childenvironment.org/factsheets/childrens_vulnerability.htm.

Child Health USA 2002. "Health Status-Infant: Infant Feeding." U.S. Department of Health and Human Resources. http://mchb.hrsa.gov/shusa02/main_pages/page_18.htm.

Children's Health Environmental Coalition. "CHEC's Safe Fish Chart." 2001–2. http://www.checnet.org/healthehouse/education/checlist-print.

Choto, R. G. "Breastfeeding: Breastmilk Banks and Human Immunodeficiency Virus." *Central African Journal of Medicine*. 36.12 (December 1990):296–300.

Colburn, Theo. *Our Stolen Future: Are We Threatening Our Fertility, Intelligence, and Survival? A Scientific Detective Study*. New York: Penguin, 1997.

Colen, Shellee. "'Like a Mother to Them': Stratified Reproduction and West Indian Childcare Workers and Employers in New York." *Conceiving the New World Order*. Berkeley and Los Angeles: U of California P, 1995. 78–102.

Commoner, Barry. "Shades of Green." *Utne Reader* (July/ August 1990): 50–63.

Congresswoman Carolyn B. Maloney. Home Page. http://www.house.gove/maloney/issues/womenchildren/breastfeeding/worldwide.htm.

"Contaminants Also Impact on Artificial Feeding." "Past Press Releases and Articles on Contaminants and Infant Feeding Available on the Baby Milk Action Website." "The Resource Centre." Update 5 October 2001. http://www.babymilkaction.org/resources/contaminants.html. (21 August 2002).

"Contaminated Milk Problem in Hawaii Nears End." *New York Times* 23 May 1982, National Desk: Section 1, Part 1, page 10.

Cook, Katsi. "Into Our Hands." *Birthing the Future*. October 1999. http://birthingthefuture.com/LivingModels/intoOurHands.html. (6/2/2003).

Corburn, Jason. "Combining Community-Based Research and Local Knowledge to Confront Asthma." *Environmental Health Perspectives* supplements 110.52 (April 2002): 241–248.

Coutsoudis, A., AE Guza, N. Rolling and H.M. Coovadis. "Free Formula Milk For Infants of HIV-Infected Women: Blessing or Curse?" Health Policy Plan. 17 (2002) 144–60.

Coutsoudis, A., "Influence of Infant-Feeding Patterns on Early Mother-to-Child Transmission of HIV-1 in Durban, South Africa: A Prospective Cohort Study." *The Lancet* 354 (August 7, 1999): 471–476.

———. "Method of Feeding and Transmission of HIV-1 from Mother to Children by 15 Months of Age: A Prospective Cohort Study." *AIDS* 15 (2001): 379–387.

Crittenden, Ann. *The Price of Motherhood: Why the Most Important Job in the World Is Still the Least Valued*. New York: Henry Holt, 2001.

Curl, Cynthia L., Richard A. Fenske et al. "Evaluation of Take-Home Organophospherus Pesticide Exposure Among Agricultural Workers and Their Children. Environmental Health Perspectives. (December 2002).

Dabis, F. V. Leroy, K. Castetbon, R. Spira, M. L. Newell, and R. Salamon. "Preventing Mother-to-Child Transmission of HIV-1 in Africa in the Year 2000." *AIDS* 14.8 (26 May 2000): 1017–1026.

Davis, Angela. "Racism, Birth Control, and Reproductive Rights." *Women, Race, and Class*. New York: Vintage Books, 1983. 202–221.

DeBarthe, Gina. "Breast-feeding and the Environment." *The Siearran* (September–October 2001). http://www.missouri.sierraclub.org/SierranOnline/SeptOct2001/16Breastmilkand%20heenvironme.html (20 September 2002).

De Bocanegra, Heike Thiel. "Breastfeeding in Immigrant Women: The Role of Social Support and Acculturation." *Hispanic Journal of Behavioral Sciences* 20.4 (November 1998): 448–468.

De LaTorre, A. and L. Rush. "The Determinants of Breastfeeding for Mexican Migrant Women. *Int Migr. Rev.* 21.3 (Fall 1987): 728–742).

Denny, Charlotte, Paul Brown, and Tim Radford. "The Shackles of Poverty." *The Guardian* (August 2002).

Dettwyler, Katherine. A. "Beauty and the Breast: The Cultural Context of Breastfeeding in the United States." *Breastfeeding: Biocultural Perspectives.* New York: Aldine de Gruyter, 1995. 167–215.

———. "Promoting Breastfeeding or Promoting Guilt?" La Leche League International 16th International Conference: Breastfeeding: Wisdom from the Past, Gold Standard for the Future. Lake Buena Vista, FL, 5 July 1999.

Diamond, Jared M. *Collapse: How Societies Choose to Fail or Succeed.* New York: Viking, 2005.

Di Chiro, Giovanna. "Local Actions, Global Visions: Remaking Environmental Expertise." *Fronteirs: A Journal of Women's Studies* 19, 2 (1998): 203—231.

"Dioxin, Incinerators, and Breastmilk." Honor the Earth: Environmental Justice and Indigenous Knowledge: Energy Policy Initiative. http://honorearth.org/ejik/energy/otherbadd/dioxin.html.

"The Diplomacy of Mothers' Milk." *New York Times* 19 May 1981, editorial desk: Sect. A, 14, Col. 1.

Dobson, Roger. "Breast Is Still Best Even When HIV Prevalence Is High, Experts Say." *BMJ* 324 (22 June 2002): 1474.

Dolk, H., M. Vrijheid, B. Armstrong et al. "Risk of Cengenital Anomalies Near Hazardous Waste Landfill Sites in Europe: The EUROHAZCON Study." *Lancet* 352 (1999): 423–427.

Drinker, Cecil K. et al., "The Problem of Possible Systemic Effects from Certain Chlorinated Hydrocarbons." *Indust. Hygiene and Toxicol.* 19 (September 1937): 283.

Dryden, John. *The Works of John Dryden,* ed. W. Scott and G. Saintsbury, XVIII, 1893.

Dwyer, A. Migrant Health Status: Profile of Population with Complex Health Problems. Migrant Clinicians Network MCN Monograph Series. Austin, TX. (1991).

Eberle, John. *A Treatise on the Disease and Physical Education of Children.* 2nd Ed. Cincinnati: Corey and Fairbank, 1834.

Ehrenreich, Barbara, and Deirdre English. *Complaints and Disorders: The Sexual Politics of Sickness.* New York: The Feminist Press, 1973.

———. *Witches, Midwives, and Nurses: A History of Women Healers.* New York: The Feminist Press.

Enfamil CD. "Smart Symphonies: Classical music to help stimulate your baby's brain development." Mead Johnson and Company. Pasadena: Disc Marketing, 2002.

Environmental Working Group. "Related News Coverage: Record Levels of Toxic Retardants Found in American Mothers' Breast Milk" www.ewg.org/news/.

Evans, Catherine. "So Why Are You Really Leaving?" *Chronicle of Higher Education* 49.35 (May 9, 2003): C4.

"FAQ on Breast Pumps." La Leche League International. 13 February 2001. http//www.lalecheleague.org/FAZ/FAQpump.html.

FASE. Foundation for Advancements in Science and Education.www.lasernet.org/.

Fergusson, D. M. et al. "Breastfeeding and Cognitive Development in the First Seven Years of Life." *Social Science & Medicine* 16(1982): 1705–1708.

Fildes, Valerie A. *Breasts, Bottles, and Babies: A History of Infant Feeding.* Edinburgh: Edinburgh UP, 1986.

————. *Wet Nursing: A History from Antiquity to the Present*. Oxford: Basil Blackwell, 1988.

Firestone, Shulamith. "The Dialectic of Sex." *Feminist Theory: A Reader*. Ed. Wendy K. Kolmar and Frances Bartkowski. London: Mayfield, 2000. 183–186.

"Fish for Life." *EPregnancy* (August 2004): 30.

Florini, Karen, and Lynn R. Goldman. "Mother's Milk Should Not Be Such a Mystery." Op-Ed. *San Francisco Chronicle*. 29 November 2000.

Flynn, John. "How Many Legs Do the Rats Have?" *Smoke Signals* (July 1993).

Fogg, Piper. "Family Time." *Chronicle of Higher Education* 49.40(6/13/2003): A10–13.

Foran, J. A., B. S. Glenn, W. Silverman. "Increased Fish Consumption May Be Risky." *JAMA: Journal of the American Medical Association*. 2.62 (July 7, 1989): 28.

Forcey, Linda Rennie. Rev. of *The Politics of Motherhood: Activist Voices from Left to Right*. By Alexis Jetter, Annelise Orleck, and Diana Taylor. *Signs* 25.1 (Autumn 1999): 301–304.

Fowler, Mary Glenn, Jeanne Bertolli, and Philip Nieburg. "When Is Breastfeeding Not Best?: The Dilemma Facing HIV-Infected Women in Resource-Poor Settings." *JAMA: Journal of the American Medical Association* 282.8 (8/25/1999): 781–784.

Fox Keller, Evelyn. "The Gender/Science System: or, Is Sex to Gender as Nature Is to Science?" *Feminism & Science*. Ed. Nancy Tuana. Bloomington: Illinois UP, 1989. 17–32.

Fraser, Antonia. *The Weaker Vessel*. New York: Vintage Books, 1985.

Fraser, Nancy. *Justice Interruptus: Critical Reflections on the 'Postsocialist' Condition*. New York: Routledge, 1997.

"Free Formula Not Advisable for HIV Say Experts." Baby Milk Action Update 31 July 2002. http://babymilkaction.org/update/update31.html.

Freed, G. L. et al. "Pediatrician Involvement in Breastfeeding Promotion: A National Study of Residents and Practitioners." *Pediatrics* 96 (1995): 490–494.

Frendenberg, Nicolas and Carol Steinsapir. "Not in Our Backyards: The Grassroots Environmental Movement." *Society and Natural Resources* 4, 235–245 (1991).

Frieden, Joyce. "Timing Is Everything in Preventing HIV Spread from Mom to Baby." WebMD Medical News Archive. 4 October 2000. http://mywebmd.com/content/article/28/1728_61990.htm.

Friend, Tim. "Cases of Cancer Decline." www.usatoday.com/news/health/cancer/2001-06-05-cancer-rates.htm.

Fuss, Diana. *Essentially Speaking: Feminism, Nature, and Difference*. New York: Routledge, 1989.

Galtry, Judith. "Extending the 'Bright Line': Feminism, Breastfeeding, and the Workplace in the United States." *Gender & Society* 14.2 (April 2000): 295–317.

————. "Suckling and Silence in the USA: The Costs and Benefits of Breastfeeding. *Feminist Economics* 3. 3 (1997): 1–24.

Garcia, A. M. "Occupational Exposure to Pesticides and Congenital Malformations: A Review of Mechanisms, Methods, and Results." *American Journal of Ind Med* 33 (1998): 232–240.

Gargan, Edward A. "Toxins Are Found in Bass in Hudson." *New York Times* 24 October 1982, late city final ed.: section 1 43.

Gartner, Lawrence. "New Breastfeeding Policy Statement from the American Academy of Pediatrics." *Breastfeeding Abstracts* 17.3 (1998): 19–20.

Gay, Jill. "Feminism, Environmental Justice, Toxic Dumps and Pesticides." *Political Environments* 7 (Fall 1999): 1–3.

Gianessi, L. P., and J. E. Anderson. "Pesticide Use in U.S. Crop Production: National Summary Report." Washington DC: National Center for Food and Agriculture Policy (1995).

Gibson, John D. "Childbearing and Childrearing: Feminists and Reform." *Virginia Law Review* 73.6 (1987):1145–1182.

Gillman, Matthew W. et al. "Risk of Overweight Among Adolescents Who Were Breastfed as Infants." *JAMA: Journal of the American Medical Association* 285.19 (16 May 2001): 2461–2467.

Giroux, Henry A. "Consuming Social Change: The 'United Colors of Benetton.'" *Cultural Critique* (Winter 1993–1994):5–31.

Golden, Janet. *A Social History of Wet Nursing in America: From Breast to Bottle.* Cambridge: Cambridge UP, 1996.

Goldsmith, Marsha F. "Specific HIV-Related Problems of Women Gain More Attention at a Price—Affecting More Women." *JAMA: Journal of the American Medical Association* 268.14 (10/14/1992): 1814–17.

"Good News on Breastfeeding Shows the Effectiveness of Environmental Campaigning." Press Release. Baby Milk Action Resources Centre. 14 May 1997. Update 5 October 2001. http://www.babymilkaction.org/resources/contamninants.html. (30 May 2002).

Gordon, "Nutritional and Cognitive Function." *Brain and Development* 19 (1997):165–170.

Greenhouse, Linda. "Justices Consider Limits of the Legal Response to Risky Behavior by Pregnant Women." *New York Times* 5 October 2000: A 26.

Griffith, John Price Crozer. *The Care of Baby: A Manual for Mothers and Nurses.* Philadelphia: Saunders, 1896.

Grinde, Donald A., Jr., and Bruce E. Johansen. *Ecocide of Native America: Environmental Destruction of Indian Lands and Peoples.* Santa Fe: Clear Light Publishers, 1995.

Gross-Loh, Christine. "Don't Trash Our Bodies!: Researching Breastmilk Toxins." *Mothering.* 122 (January/February 2004): 54–65.

Grossman, Karl. "The People of Color Environmental Summit." *Unequal Protection: Environmental Justice and Communities of Color.* Ed. Robert D. Bullard. San Francisco: Sierra Club Books, 1994. 272–297.

Guay, Laura A., and Andrea J. Ruff. "HIV and Infant Feeding—An Ongoing Challenge." *JAMA: Journal of the American Medical Association* 286.19 (November 21, 2001):2462–2465.

Haller, Mark H. *Eugenics: Hereditarian Attitudes in American Thought.* New Brunswick: Rutgers UP, 1963.

Hamilton, Cynthia. "Concerned Citizens of South Central Los Angeles." *Unequal Protection: Environmental Justice and Communities of Color.* Robert D. Bullard, ed., San Francisco: Sierra Club Books, 1994, 207–219.

Hanley, Robert. "Jersey Citing PCB Levels, Urges Limit on Eating Five Kinds of Fish." *New York Times* 14 December 1982, late city final ed.: B 1.

———. "New Jersey Journal." *New York Times* 21 September 1980, late city final ed.: sect. 11 3.

Harding, Sandra. *Feminism & Science.* Ed. Nancy Tuana. Bloomington: Illinois UP, 1989. 17–31.

Harrison, Kathryn. "Too Close to Home: Dioxin Contamination of Breast Milk and the Political Agenda." *Policy Sciences* 23 (2001):35–62.

Hartmann, Betsy. "Population Control and Foreign Policy." *Covert Action* 39 (Winter 1991–1992).

Hausman, Bernice L. *Mother's Milk: Breastfeeding Controversies in American Culture.* New York: Routledge, 2003.

Hawken, Paul. "Corporatization of the Commons." University of Colorado, Boulder 10 March 2001. Alternative Radio.

"Hazards of New York Fin Fish." *New York Times* 2 June 1982, late city final ed.: C8.

"Health Claims, Medical Foods and '6 Months' in the Spotlight at Codex." Baby Milk Action Update 30. December 2001. http://www.babymilkaction.org/update/update30.html.

"Health Effects: Toxics." Greenpeace USA. http://www.greenpeaceusa.org/toxics/impactstext.htm. (20 September 2002).

"Healthy Milk, Healthy Baby: Chemicals in Mother's Milk." Natural Resources Defense Council. 22 May 2001. http//www.nrdc.org/breastmilk/envpoll.asp (25 September 2001).

HHS Blueprint for Action on Breastfeeding, Washington, DC: Department of Health and Human Services, Office of Women's Health, 2000.

High PCB Levels Found in Eskimo Breastmilk." *New York Times* 7 February 1989, late city final ed.: Science Desk C 9.

"HIV and Breastfeeding—Unethical Research on Maternal Deaths?" Babymilk Action. June 2001. http://www.babymilkaction.org/update29.html.

"HIV and Infant Feeding." Implementation of Guidelines: A Report of the UNICEF-UNAIDS-WHO technical Consultation on HIV and Infant Feeding." WHO: Joint United Nations Programme on HIV/AIDS (UNAIDS). Geneva. 20–22 April 1998.

"HIV and Infant Feeding: New UN Policies Raise Critical Issues." Infact Canada. Summer 98. http://www.infanctcanada.ca/hiv.htm.

Hobson, Geary, ed. *The Remembered Earth.* Albuqerque, NM: Red Earth Press, 1979.

Hornblower, Margot. "Firms Exporting Products Banned as Risks in U.S.; U.S. Firms Export Products Banned Here as Health Risks." *Washington Post* 25 February 1980, final ed.: first section A1.

Horowitz, Julie Bouche. "A Special Gift: Breastfeeding an Adopted Baby." *Mothering* 104 (January/ February 2001): 62–69.

Horwood, L. J., and D. M. Fergusson. "Breastfeeding and Later Cognitive and Academic Outcomes." *Pediatrics* 101(1998):E9.

Hubbard, Ruth. *Feminism & Science.* Ed. Nancy Tuana. Bloomington: Illinois UP, 1989. 119–131.

Huber, Peter. *Hard Green: Saving the Environment form the Environmentalists: A Conservative Manifesto.* New York: Basic Books, 1999.

Hulse, Carl. "Senate Outlaws Injury to Fetus During a Crime." *New York Times* 26 March, 2004: A1, 16.

"Human Health Studies." http://www.albany.edu/sph/superfund/hhhealth.html. (6/2/2003).

Human Milk Banking Association of North America. http://www.hmbana.org/processing.html. (April 10, 2004.)

Humphrey, Harold E. B. et al. "PCB Congener Profile in the Serum of Humans Consuming Great Lakes Fish." *Environmental Health Perspectives* 108.2 (February 2000).

"Indigenous Peoples and POPS: Briefing Paper in Preparation for the UNEP POPs Intergovernmental Negotiating Committee (INC4) Meeting, in Bonn, Germany, 20–25 March, 2000." Indigenous Environmental Network. *EJ Times: The Environmental Justice Newsletter from Sierra Club* 1.3 (April–June 2000): 10–12.

INFACT Canada. "Breastmilk: the Perfect Renewable Resource." The Ecology of Breastfeeding and the Dangers of Artificial Feeding. Spring, 1997 http://www.infactcanada.ca/ren_res.htm.

Institute of Medicine of the National Academies. "Projects." Workshop #1: Rebuilding the Unity of Health and the Environment: A New Vision of Environmental Health for the 21st Century. www.iom.edu.

"Is Mother's Milk Fit For Human Consumption?" EDF Advertisement. *New York Times* 29 March, 1970. http://www.environmentaldefense.org/documents/244_Where%20All%Beagan.htm. (20 September 2002).

Jacobson, J. L., and S. W. Jacobson. "Effects of In Utero Exposure to PCBs and Related Contaminants on Cognitive Function in Young Children." *Pediatrics* 116 (1990): 38–45.

———. "Intellectual Impairment in Children Exposed to PCBs in Utero." *New England Journal of Medicine* 335 (1996): 783–789.

James, Rebecca. "Cornell Author Warns About Breast Milk: A Lifetime of Toxins Concentrates in Breast Milk Says Breast-Feeding Author." *Post-Standard* 22 April 2000, final ed.: A1.

Jeans, Philip C. "A Review of the Literature on Syphilis in Infancy and Childhood, Part 1." *American Journal of Diseases of Children* 20 (1920).

Jelliffe, Derrick B., and E. F. Patrice Jelliffe. *Human Milk in the Modern World: Psychosocial, Nutritional, and Economic Significance.* Oxford: Oxford UP, 1978.

John, A. M., and R. Martorell. "Incidence and Duration of Breast-feeding in Mexican-American Infants, 1970–1982." *American Journal of Clinical Nutrition* 50 (1989): 868–874.

John, G. C., B. A. Richardson, R. W. Nduati, D. Mbori-Ngacha, and J. K. Kreiss. "Timing of Breastmilk HIV-1 Transmission: A Meta-Analysis." *East African Medical Journal.* 78.2 (February 2001): 75–79.

Kamerman, Sheila B., Alfred J. Kahn, and Paul Kingston. *Maternal Policies and Working Women.* New York: Columbia UP, 1983.

Keating, Martha. "Mercury, Power Plants and the Fish We Eat." June 2000. Clean Air Task Force. Available at (617) 292-0234, Fax: (617) 292-4933.

Kelly, Katsy. "Milk Worries." *US News and World Report.* (6 October 2003). http://ewg.org/news/story.php?id=2047.

Kent, George. "MOMM on Trial: The Missing Testimony on Human Rights, Breastfeeding, and Forced Drug Treatment." MOMM: Mothers Opposing Mandatory Medicine. 23 April 1999. http://www.informedmomm.com/stories/mom-on-tiral.htm.

Killingsworth, Jimmie M., and Jacqueline S. Palmer. "The Discourse of 'Environmentalist Hysteria.'" *The Quarterly Journal of Speech.* 81.1 (February 1995): 1–19.

King, Ynestra. "The Ecology of Feminism and the Feminism of Ecology." *Healing the Wounds: The Promise of Ecofeminism.* Ed. Judith Plang. Philadelphia: New Society Publishers, 1989. 18–29.

Kirk, Gwyn. "Ecofeminism and Environmental Justice: Bridges Across Gender, Race, and Class." *Frontiers:* (1997) 28.2 2–20.

Klemesrud, Judy, and Michelle Dooley. "Toxic Risks to Babies Underlined." *New York Times* 25 September, 1980, Late City final ed.: C1, Col. 1.

Krauss, Celene. "Women and Toxic Waste Protests: Race, Class and Gender as Resources of Resistance." *Qualitative Sociology.* 16. 3 (1993): 247–260.

Kristeva, Julia. "Woman Can Never Be Defined." *New French Feminisms: An Anthology.* Ed. Elaine Marks and Isabelle de Courtivron. New York: Schocken Books, 1980. 137–141.

Kuhn, Brooke. "Should a Mother with HIV Breastfeed Her Baby?" WebMD Medical News Archive. WebMDHealth. 16 November 1999. http://my.webmd.com/content/article/19/1728_50598.htm?

La Duke, Winona. *All Our Relations: Native Struggles for Land and Life.* Cambridge, MA: South End Press, 1999.

———. Interview with Katsi Cook. Akwasasne Articles: On Breastmilk, PCBs, and Motherhood. http://www..honorearth.com/akwasasne4.html. (November 5, 1999).

Landrigan, P. J., and J. E. Carlson. "The Future of Children." *Environmental Policy and Children's Health* 5 (1995): 34–52.

Lanting, C. I. et al. "Breastfeeding and Neurological Outcome at 42 months." *Acta Paediatrica* 87 (1998)1224–1229.

———. "Neurological Differences Between 9-Year Old Children Fed Breast-Milk or Formula-Milk as Babies." *Lancet* 344(1994):1319–1322).

Laug, Edwin P., Frieda M. Kunze, and C. S. Prickett. "Occurrence of DDT in Human Fat and Milk." *A.M.A. Archives of Industrial Indus. Hygiene and Occupat. Med.* 3 (1951): 245–246.

Lavelle, M., and M. Coyle. "Unequal Protection: The Racial Divide in Environmental Protection." *Nat Law J* Special Issue 9.5 (1992).

Law, Jules. "The Politics of Breastfeeding: Assessing Risk, Dividing Labor." *Signs* 25.2 (Winter 2000): 407–439.

Lawrence R. A. "A Review of the Medical Benefits and Contraindications to Breastfeeding in the United States." *Maternal and Child Health Technical Information Bulletin* (Arlington, Va: National Center for Education in Maternal and Child Health, 1997):4.

———. *Breastfeeding : A Guide for the Medical Profession.* 4th ed. St. Louis: Mosby, 1994.

Layne, Linda L. "Breaking the Silence: An Agenda for a Feminist Discourse of Pregnancy Loss." *Feminist Studies* 23.2 (1997): 289–318.

———. "In Search of Community: Tales of Pregnancy Loss in Three Toxically Assaulted U.S. Communities." *Women's Studies Quarterly* 1 & 2 (2001): 25–50.

Lee, Charles. "Environmental Justice: Building A Unified Vision of Health and the Environment." *Environmental Health Perspectives Supplements* 110. Issue Suppl.2 (April 2002): 141–145.

Lefferts, Lisa. "Something Fishy: Fish May Be Better for You Than Red Meat, But It's Not Perfect." *Everyday Magazine. St. Louis Post-Dispatch* 11 February 1989, five star ed.: 1D.

Lewin, Tamar. "Study Links Working Mothers to Slower Learning." The *New York Times,* July 17, 2000, A14.

LHMU. "Child Care Work Still Pays Little in the U.S." 14 March 2002. http://lhmu.org.au/childcare/news/22.html. (28 May 2002).

Lobet, Ingrid. "Your Chemical Body Burden." *Living On Earth*. Natl. Public Radio. 16 January 2004.

Lohr, Steve. "Is Wal-Mart Good for America?" Week in Review. *New York Times* 7 December 2003, late ed, final: 4.1.

Lorde, Audre. "Age, Race, Class, and Sex: Women Redefining Difference." *Sister Outsider: Essays and Speeches by Audre Lorde*. Freedom, CA: The Crossing Press, 1984. 114–123.

———. "The Uses of Anger: Women Responding to Racism." *Sister Outsider: Essays and Speeches by Audre Lorde*. Freedom, CA: The Crossing Press, 1984. 124–133.

Love, Spencie. *One Blood: The Death and Resurrection of Charles R. Drew*. Chapel Hill: U of North Carolina P, 1996.

Lucas A., Morley R. et al. "Breastmilk and Subsequent Intelligence Quotient in Children Born Preterm" 339 (1992): 261–264.

———. "Early Diet in Preterm Babies and Developmental Status at 18 Months." *Lancet* 335 (1990): 1477–1481.

Lyman, Francesca. "Concerns Raised on Breast milk." *MSNBC Home*. http://www.msnbc.com/news/227752.asp.

Lyons, Gwynne. "Chemical Trespass: A Toxic Legacy." Executive Summary: A WWF-UK Report. June 1999. 1–18.

Maher, Vanessa, ed. "Breast-Feeding in Cross-Cultural Perspective: Paradoxes and Proposals." *The Anthropology of Breast-Feeding: Natural Law or Social Construct*. Oxford: Berg, 1992.

Malveaux, Julianne. "MS. News." MS. (October/ November 2001): 24.

Martin, Kallen M. "Adwesasne Environments, 1999: Relicensing a Seaway After a Legacy of Destruction." 1999. http://nativeamericas.aip.cornell.edu/spr99features/martinspr99.html. (8/17/2001).

Martin, P., R. Mines, and A. Diaz. "A Profile of California Farmworkers." *California Agriculture* 6 (1985): 16–18.

Martyn, Tessa. "Infant Feeding and HIV: Saviours or Culprits? HIV, Infant Feeding, and Commercial Interests." Baby Milk Action Resources Centre. April 2001. http: //www.babymilkaction.org/resources/hiv01.html.

Maxa, Rudy. "Did We Forget the Deadly Message of Silent Spring?" *Washington Post Magazine*. 30 May 1982, final ed.: front page.

Mc Intosh. Peggy. "White Privilege and Male Privilege: A Personal Account of Coming to See Correspondences Through Work in Women's Studies." *Race, Class, and Gender: An Anthology*. Ed. Margaret Andersen and Patricia Hill Collins. 4th Ed. Belmont, CA: Wadsworth, 2001. 95–105.

McMahon, Martha. *Engendering Motherhood: Identity and Self-Transformation in Women's Lives*. New York: Guilford Press, 1995.

Merchant, Carolyn. *The Death of Nature: Women, Ecology and the Scientific Revolution*. San Francisco: Harper Collins, 1980.

———. *Earthcare: Women and the Environment*. New York: Routledge, 1995.

Merrill, "Learning How to Mother: An Ethnographic Investigation of an Urban Breastfeeding Group." *Anthropology & Education Quarterly* 18.3 (September 1987): 222–239.

Mies, Maria, and Vandana Shiva. *Ecofeminism*. Halifax, Nova Scotia: Fernwood Publications, 1993.

Mies, Maria. "Liberating the Consumer." *Ecofeminism*. Maria Mies and Vandana Shiva. Halifax, Nova Scotia: Fernwood Publications, 1993.

———. "The Myth of Catching Up Development." *Ecofeminism*. Maria Mies and Vandana Shira. Halifax, Nova Scotia: Fernwood Pubications, 1993.

"Milk Money." *20/20*. ABC News. 4 June 2004.

Miller, N. H, D. J. Miller, and M. Chism. "Breastfeeding Practices among Resident Physicians." *Pediatrics* 98.3 part 1 (September 1996): 434–47.

Miotti, Paolo G. et al. "HIV Transmission Through Breastfeeding: A Study in Malawi." *JAMA: Journal of the American Medical Association* 282.8 (25 August 1999): 744–50.

Mitchell, Alison. "Senate Approves Nuclear Waste Site in Nevada Mountain." *New York Times* 10 July 2002, late ed.: A14.

Moraga, Cherrie. *Heroes and Saints and Other Plays*. Albuquerque, NM: South End Press, 1994.

Mortensen, Erik Lykke et al. "The Association Between Duration of Breastfeeding and Adult Intelligence." *JAMA: Journal of the American Medical Association* 287.18 (8 May 2002): 2365–2346.

Narayan, Uma. *Dislocating Cultures: Identities, Traditions, and Third World Feminism*. New York: Routledge, 1997.

National Academy of Sciences. "Pesticides in the Diets of Infants and Children." Washington, DC: National Academy Press, 1993.

National Research Council. "Pesticides in the Diets of Infants and Young Children." Washington, DC: National Academy Press, 1993.

Nduati, R. et al. "Effect of Breastfeeding and Formula Feeding on Transmission of HIV-1: A Randomized Clinical Trial." *JAMA: The Journal of the American Medical Association* 283.9 (1 March 200): 1167–1174.

"Nestle Uses HIV to Push Infant Formula in Africa in Battle with Wyeth." Babymilk Action Update 30. December 2001. http://www.babymilkaction.org/update/update30.html.

"No Alarm Over Dioxin in Human Milk." *The Record* 18 April 1986: A12.

Norwood, Christopher. *At Highest Risk: Environmental Hazards to Young and Unborn Children*. New York: McGraw-Hill, 1980.

NOW: National Organization for Women. Home Page. "Now Appreciates AAP Recommendations on Breast Feeding, Calls on Business and Society to Support Findings." 3 December 1997. Press Release. http://www.now.org/press/12–97/12–03–97.html.

———. "NOW Demands Greater Acceptance and Access for Breastfeeding Mothers." http://www.now.org/nnt/05-98/breastfd.html.

Olshan A. F., and E. M. Faustman. "Male-Mediated Developmental Toxicity." *Annual Review of Public Health* 14 (1993): 159–181.

Ortiz, A. et al. "Because They Were Born From Me: Negotiating Women's Rights in Mexico." *Negotiating Reproductive Rights: Women's Perspectives Across Countries and Cultures*. Eds. R. Petchesky and K. Judd. London, UK: Zed Books, Ltd., 1998.

"Our Story." About Us. http://www.checnet.org/about_us/our_story.html (16 June 2002).

Palmer, Gabrielle. *The Politics of Breastfeeding*. London: Pandora, 1988.

Pardo, Mary. *Mexican American Women Activists: Identity and Resistance in Two Los Angeles Communities*. Philadelphia: Temple UP, 1998.

———. "Identity and Resistance: Mexican American Women and Grassroots Activism in Two Los Angeles Communities." PhD. diss. University of California, Los Angeles, 1990.

Patadin, S. et al. "Effects of Environmental Exposure to PCBs and Dioxins on Cognitive Abilities in Dutch Children at 42 Months of Age." *Journal of Pediatrics* 134 (1999): 33–41.

Patton, Sharyle. "Update on Biomonitoring Legislative Proposal (Bill 689 Senator Ortiz)." Health and Environment Program Commonweal. 19 March, 2003:1–7.

"PCB Threat to Infants Cited." *St. Petersburg Times* 15 April 1988, city ed.: National Medicine 9A.

Perfecto, Ivette. "Pesticide Exposure of Farm Workers and the International Connection." *Race and the Incidence of Environmental Hazards: A Time for Discourse.* Ed. Bunyan Bryant and Paul Mohai. Boulder: Westview Press, 1992. 177–203.

Pesticide Action Network International. "Farm Worker Exposure." www.pan-international.org/

Petersen, Melody. "Advertising: Breastfeeding Ads Delayed by a Dispute Over Content." *New York Times.* 4 December 2003. http://nytimes.com/2003/12/04/business/media/04adcol.html.

Pew Environmental Health Commission. *Healthy from the Start: Why America Needs a Better System to Track and Understand Birth Defects and the Environment.* Baltimore: John Hopkins School of Public Health, 1999.

Pharr, Suzanne. *Homophobia: A Weapon of Sexism.* Expanded Edition. Berkeley: Chandon Press, 1997. 53–64.

"Philanthropy Online." Babymilkaction. July 1999. http://babymilkaction.org/update/update 25.html.

Phillips D. P., E. J. Kanter, B. Bednarczyk, et al. "The Importance of the Lay Press in the Transmission of Medical Knowledge to the Scientific Community." *New England Journal of Medicine* 325 (1991): 1180–1183.

Pillay, K., E. Spooner, L. Kahn, H. M. Coradin. "The Courageous American 'No' to W.H.O.'s Infant-Formula Code." *New York Times* 24 May 1981, editorial desk: Sect. 4, page 18, col. 4.

"Physicians and Breastfeeding Promotion in the United States: A Call For Action." *Pediatrics* 107.3 (March 2001): 584–589.

Platt, Kamala. "Chicana Strategies for Success and Survival: Cultural Poetics of Environment Justice from the Mothers of East Los Angeles." *Frontiers: A Journal of Women's Studies* 18.2 (1997): 48–72.

Platypus Media. "Breastfeeding Rates." 2001. http://www.platypusmedia.com/adult_rates.html.

Pohl, Otto. "European Environmental Rules Propel Change in U.S." *New York Times.* (6 July 2004). http://ewg.org/news/story.php?id=2767.

Pollock, Linda A. *Forgotten Children: Parent-Child Relationships from 1500 to 1900.* Cambridge: Cambridge UP, 1983.

"Population and Consumption." Population and Consumption—Global Population and Environment. Sierra Club. 2004. http://www.sierraclub.org/population/consumption/.

"Position Papers: NY Times ADS." Center for Children's Health and the Environment. 1 June, 2002. http://www.choldenvironment.org/position.htm.

"President of Botswana Meets NGOs." BabymilkAction. June 2001. http://www.babymilkaction.org/update/update29.html.

"Prevention and Early Detection." *Washington Post* 14 August 1985, final ed. "Health":7.

Prindeville, Diane-Michelle, and John G. Bretting. "Indigenous Women Activists and Political Participation: The Case of Environmental Justice." *Women & Politics* 19.1 (1998): 39–58.

Prior, Mary, ed. *Women in English Society 1500–1800*. London: Methuen, 1985.

Pryse, Marjorie. "Defining Women's Studies Scholarship: A Statement of the National Women's Studies Association Task Force on Faculty Roles and Rewards." NWSA June 1999.

Pulido, Laura. "Community, Place, and Identity." *Thresholds in Feminist Geography: Difference, Methodology, Representation*. Ed. John Paul Jones III, Heidi J. Nast, and Susan M. Roberts. Lanham, NY: Rowman and Littlefield, 1998.

Quinn, Daniel. *Ishmael: An Adventure of the Mind and Spirit*. New York: Bantam, 1992.

Radford, Andrew. "The Ecological Impact of Bottle Feeding." Babymilk Action. 1991. http://home.clara.net/abm/subpages/eco_impact.htm. (31 May 2002).

Reed, C. "A Study of the Conditions that Require the Removal of the Child from the Breast." *Surg Gynecol Obstet* 6 (1908):514–527.

Regenstein, Lewis. "Poisons That Are Boomeranged Out of the U.S." *New York Times* 25 March 1983, late city final ed.: editorial desk sect. A 14.

"Report on the Mt. Olive Pickle Boycott by the Farm Labor Organizing Committee and Migrant Farm Worker Conditions in North Carolina and in the United States." Farm Worker Conditions. http://www.nccusa.org/publicwitness/mtolive/conditions.html. (12/9/2003).

"Researchers Look at Contaminants in Human Milk." http://three.parilist.net/mailman/listinfo/checnet-forum. (18 April 2003).

Richards, Dave. *Seeing Through the Spin: Public Relations in the Globalised Economy*. United Kingdom: Baby Milk Action, 2001.

Roediger, David. *Towards the Abolition of Whiteness*. London: Verso, 1994.

Rogan, W. J. "Pollutants in Breast Milk." *Archives of Pediatric and Adolescent Medicine* 150 (1996):981–990.

Rogan, W. J., P. J. Blanton, C. J. Portier, and E. Stallard. "Should the Presence of Carcinogens in Breast Milk Discourage Breast Feeding?" *Regulatory Toxicology and Pharmacology* 13 (1991):228–240.

Rogan, W. J., and Gladen, B. C. "Study of Human Lactation for Effects of Environmental Contaminants: The North Carolina Breast Milk and Formula Project and Some Other Ideas." *Environmental Health Perspectives* 60 (1985):215–221.

Romero-Gwynn, Eunice, and Lucia Carias. "Breast-Feeding Intentions and Practice Among Hispanic Mothers in Southern California." *Pediatrics* 84.4 (October 1989): 626–632.

Rosenblum, Karen E., and Toni-Michelle C. Travis. *The Meaning of Difference: American Constructions of Race, Sex and Gender, Social Class, and Sexual Orientation*. 2nd ed. Boston: McGraw-Hill, 2000.

Ruowei, Li; L. Grummer-Strawn, Z. Zhao L. Barker and A. Mokdad. "Prevalence of Breastfeeding in the United States: The National Immunization Survey." *Pediatrics*. 111 (2003): 1198–1201.

Russo, Ann. "We Cannot Live Without Our Lives: White Women, Antiracism, and Feminism." *Third World Women and the Politics of Feminism.* Ed. Chandra T. Mohanty, Ann Russo, Lourdes Torres. Indiana: Indiana UP, (1991): 297–313.

Sager, Mike, and B. D. Colen. "Scientist Says Fear May Prompt Complaints on PCB." *Washington Post.* 10 November 1999, Metro: C3.

Savitz, D. A., T. Arbuckle, D. Kaczor, and K. M. Curtis. "Male Pesticide Exposure and Pregnancy Outcome." *American Journal of Epidemiology* 146. 121 (1997): 1025–1036.

SB 1168 Senate Bill—Bill Analysis. 22 June 2004. http://info.sen.ca.gov/pub/bill/sen/sb_1151-1200/sb_ 1168_cfa_20040621_115230_asm_comm.html.

"Scare Stories Again." "Past press releases and articles on contaminants and infant feeding available on the Baby Milk Action website." Baby Milk Action Resources Centre. December 1999. Update 5 October 2001. http://www.babymilkaction.org/resources/contamninants.html. (21 August 2002).

Schafer, Kristin. "Biomonitoring: A Tool Whose Time Has Come. Finding Pesticides in Our Bodies." *Global Pesticide Campaigner* 14.1 (April 2004). www.panna.org.

Schanler, R. J. et al. "Pediatrician Involvement in Breastfeeding Promotion: A National Study of Residents and Practitioners." *Pediatrics* 103.3 (1999): E 35. http://www.pediatrics.org/cgi/content/full/103/3/e35.

Schell, L. M. and A. M. Tarbell. "A Partnership Study of PCBs and the Health of Mohawk Youth: Lessons from Our Past and Guidelines to Our Future." *Environmental Health Perspectives* 106.3 (1998): 833–840.

Schettler, Ted, Gina Solomon, Maria Valenti, and Annette Huddle. *Generations at Risk: Reproductive Health and the Environment.* Cambridge: MIT Press, 1999.

Schmidt, Karen F. "Preventing Fires, Igniting Questions." *US News and World Report.* (April 26 2004). http://ewg.org/news/story.php?id=2455.

Schmied, Virginia, and Deborah Lupton. "Blurring the Boundaries: Breastfeeding and Maternal Subjectivity." *Sociology of Health & Illness* 23.2 (2001): 234–250.

Schneider, Keith. "Efforts Revive River But Not Mohawk Life." *New York Times* 6 June 1994, late ed.:final B.

———. "Fetal Harm Is Cited as Primary Hazard in Dioxin Exposure." *New York Times* 11 May 1994, late ed.: final section 1, 1.

Schreiber, Judith S. "Parents Worried About Breast Milk Contamination: What Is Best For Baby?" *Children's Environmental Health* 48.5 (October 2001): 1113–1127.

———. "Prediced Infant Exposure to Tetrachoroethylene in Human Breastmilk." *Risk Analysis* 13.5 (1993): 515–24.

"Science Watch: A Really Big Syzygy." *New York Times* 31 March 1981, late city final ed.: C 2.

Seager, Joni. *Earth Follies: Coming to Feminist Terms with the Global Environmental Crisis.* New York: Routledge, 1993.

———. "Rethinking the Environment: Women and Pollution." *Political Environments* 3 (Spring 1996): 1–2. http://www.cwpe.org/issues/environment_html/joni.html.

Seely, Hart. "Toxins Remain 18 Years Later." 2001. http://envirocrime.com/forum/_disc5/0000000b.htm. (9/25/2002).

Sengupta, Somini. "A Toxic Dump and a Legacy of Sickness: Mohawks and G.M. At Odds Over Cleanup." *New York Times* 7 April 2001, late ed: B1.

Sexton, K., and Y. B. Anderson, eds. "Equity in Environmental Health: Research Issues and Needs." *Toxicol Ind Health* (Special Issue) 9 5 (1993):679–977.

Shabecoff, Philip. "Dioxin in Breast Milk Is Evaluated in Private Study." *New York Times* 18 December 1987, Late City final ed.: A17.

Shaftel, Norman. "A History of the Purification of Milk in New York: or How Now Brown Cow." *Sickness and Health in America: Readings in the History of Medicine and Public Health*. Ed. Judith Walzer Leavitt and Ronald L. Numbers. 3rd Edition. Madison: U of Wisconsin P, 1997.

Shepard, Peggy M., Mary E. Northridge, Swati Praknah, Gabriel Stover. "Preface: Advancing Environmental Justice through Community-Based Participatory Research." *Environmental Health Perspectives Supplements* 110.52 (April 2002). http://ehpnieh5.gov/docs/2002/suppl-2/toc.html.

Ship, Susan Judith. "Environmental Contaminants and Traditional Foods." http://www.niichro.com/Environ/Enviro2.html.

Shipley, Sara. "EPA Will Phase Out Flame Retardant Chemical Found in Breast Milk." 7 November 2003, www.ewg.org/news/story. *St. Louis Dispatch*.

Shiva, Vandana. *Staying Alive: Women, Ecology & Development*. London: Zed Books, 1989.

———. "The Impoverishment of the Environment" *Ecofeminism*. Maria Mies and Vandana Shira. Halifax, Nova Scotia. Fernwood Publications, 1993.

Silbergeld, Ellen K. "Dioxin in Breast Milk Raises New Health Concerns." *EDF Column*. XIXX, 2 (April 1988). http://www.edf.org/pubs/EDF-Letter/1988/Apr/ j _dioxin.html.

Sinclair, Marti. "Environmental Justice and POPs." *EJTimes: The Environmental Justice Newsletter from Sierra Club*. 1.3 (April–June 2000): 3–4.

Smith, Andy. "Ecofeminism Through an Anti-Colonial Framework." *Ecofeminism: Women, Culture, Nature*. Ed. Karen J. Warren. Bloomington: Indiana UP, 1997. 21–37.

Smith, Barbara. "Racism and Women's Studies." *But Some of Us Are Brave*. Old Westbury: Feminist Press, 1982.

Smith, F. B. *The People's Health 1830–1916*. London: Croom Helm, 1979.

Smith-Nonini, Sandy. "Uprooting Justice—A Report on Working Condition for North Carolina Farm Workers and The Farm Labor Organizing Committee's Mt. Olive Initiative." Institute for Southern Studies, Durham, N.C. www.ncccu.org/ publicwitness/mtolive/boycott.html.

Snell, Marilyn Berlin. "Interview: Theo Colborn." *MotherJones.com*. March/ April 1998. http://www.motherjones.com/news/qa/1998/03/snell.html.

Snyder, J. Ross. "The Breast Milk Problem." *JAMA: Journal of the American Medical Association* 51 (1908):1213.

Solomon, Stephen. "The Controversy Over Infant Formula." *New York Times*. 6 Dec. 1981, late city final ed.: 6, 92, col. 1.

Spanier, Bonnie. *Im/Partial Science: Gender Ideology in Molecular Biology*. Bloomington: Indiana UP, 1995.

"Spanish CHEC Lists Now Available." CHEC Health—eNews. 6 May, 2004. http:/ /www.checnet.org/healthehous/education/articles-detail.asp?Main_ED=854.

Spencer, Cindie. "Feminist Sitings in a 'Toxic Culture': Linking Theory with Practice." National Women's Studies Conference. Minneapolis, MN. June 2001.

Steele, Ann. "When to Procreate." *Chronicle of Higher Education* 49.44 (11 July/ 2003): C4.

Steinfeld, Melvin. *Our Racist Presidents.* San Romano, CA: Consensus Publishers, 1972.

Steingraber, Sandra. *Having Faith: An Ecologist's Journey to Motherhood.* Cambridge: Perseus Publishing, 2001.

———. *Living Downstream: A Scientists's Personal Investigation of Cancer and the Environment.* New York: Vintage Books, 1997.

———. "Protecting the First Environment: Ecological Threats to Women, Pregnancy, and Breast Milk." Talk. Siena College, Albany, New York. 3 April 2003.

Stevens, Susan. "Home Toxic Home: Common Products Can Make Living in Your Home Hazardous to Your Health." *Chicago Daily Herald* 6 June 2005. www.ewg.org/news/story.

Stewart, P. "PCBs/Methylmercury: The Oswego Study." Reported at Childeren's Health and the Environment: Mechanisms and Consequences of Developmental Neurotoxicology. Little Rock, AR. October 1999.

"Study of Chemical Stirs Breast-Feeding Concerns." *Seattle Post-Intelligencer* 30 October 1990, final ed.:A2.

Sucher, Lauren, and Bill Walker. "Toxic Fire Retardants Found in U.S. Women's Breastmilk: Study Urges Chemical Ban, But Says Breastfeeding Still Best for Baby & Mom." *Environmental Working Group: Mother's Milk: Record Levels of Toxic Fire Retardants Found in American Mothers' Breast Milk.* 23 September 2003. http://www/ewg.org/reports/mothersmilk/release_20030923.php.

Surgeon General. Surgeon General's Workshop on Breastfeeding and Human Lactation. Presented by the U.S. Department of Health and Human Services, Public Health Service. Rochester, NY, 1984.

"Swazi Health Minister Comments on HIV." Babymilk Action Update 28. November 2000. http://www.babymilkaction.org/update/update28.html.

Taha, Taha E. et al. "Nevirapine and Zidovudine at Birth to Reduce Perinatal Transmission of HIV in an African Setting: A Randomized Controlled Trial." *JAMA: Journal of the American Medical Association* 292.2 (14 July 2002): 202–210.

Tallman, Valerie. "The Toxic Waste of Indian Lives." *Covert Action* 17 (Spring 1992).

Taub, Nadine. "From Parental Leaves to Nurturing Leaves." *New York University Review of Law and Social Change* 13.2 (1985): 381–405.

Taub, Nadine, and Wendy Williams. "Will Equality Require More than Assimilation, Accomodation or Separation form the Existing Social Structure?" *Rutgers Law Review* 37 (1985):825–844.

Thomas, Katie. "Toxic Threats to Tribal Lands." *We Have Many Voices* 1.5 (31 March 2001): 1–5. http://www.turtletrack.org/ManyVoices/Issue_5/ Toxic_Land.htm.

Ticknor, Caleb B. *A Guide for Mothers and Nurses in the Management of Young Children.* New York: Dodd and Taylor, 1939.

Umansky, Lauri. *Motherhood Reconceived: Feminism and the Legacy of the Sixties.* New York: New York UP, 1996.

UNICEF. "Nutrition." "Breastfeeding League." www.unicef.org/pon96/nubreast.htm.

United Church of Christ Commission for Racial Justice. *Toxic Wastes and Race in the United States: A National Study of the Racial and Socioeconomic Characteristics of Communities With Hazardous Waste Sites.* New York: United Church of Christ Commission for Racial Justice, 1987.

U.S. Department of Agriculture. *USDA Statistics, 1986.* Washington, DC: U.S. Government Printing Office, 1986.

U.S. Department of Health and Human Services. Report of the Secretary's Task Force on Black and Minority Health. Washington, DC: DHHS, 1985.

U.S. Environmental Protection Agency. "Environmental Equity: Reducing Risk for All Communities." Report to the Administrator. EPA 230-R-92-008. Washington, DC: U.S. Environmental Protection Agency, 1992.

———. "Fish Advisories." 3 August, 2004. http://www.epa.gov/ost/fish/

———. "Health Assessment Document for 2,3,7,8—Tetrachlorodibenzo-p-dioxin (TCDD) and Related Compounds." External Review Draft. 1994.

———. *Pesticides Fact Book.* Washington, DC: EPA Office of Public Affairs. 1986.

Uram, Eric. "The Fox River's Legacy: New Settlers in the Region." *EJTimes: The Environmental Justice Newsletter from Sierra Club* 1.3 (April–June 2000).

Van de Perre, P. "Breastmilk Transmission of HIV-1: Laboratory and Clinical Studies." *Annals of the New York Academy of Sciences.* 918 (November 2000) 122–127.

Van Esterick, Penny. *Beyond the Breast-Bottle Controversy.* New Brunswick, NJ: Rutgers UP, 1989.

———. "Breastfeeding and Feminism." *International Journal of Gynecology and Obstetrics* 47 Supplement (1994):S41–S54.

———. "Lessons From Our Lives: Breastfeeding in a Personal Context." *Journal of Human Lactation* 10.2 (1994):71–74.

———. "Thank You, Breasts! Breastfeeding as a Feminist Issue." *Ethnographic Feminisms: Essays in Anthropology.* Ed. Sally Cole and Lynne Phillips. Ottowa: Carleton UP, 1995. 75–99.

Vogel, Lise. "Debating the Difference: Feminism, Pregnancy, and the Workplace. *Feminist Studies* 16.1 (1990): 9–32.

———. *Mothers on the Job: Maternity Policy in the U.S. Workplace.* New Brunswick, NJ: Rutgers UP, 1993.

Walker, Jerome. *The First Baby: His Trials and the Trials of His Parents.* New York: Brown & Derby, 1881.

Walkowiak, J. et al. "Environmental Exposure to Polychlorinated Biphenyls and Quality of the Home Environment: Effects on Psychodevelopment in Early Childhood." *Lancet* 358 (2001): 1602–1607.

Wambach, K. A., and C. Cole. "Breastfeeding and Adolescents." *Journal of Obstetrics, Gynecologic, and Neonatal Nursing: JOGNN/NAACOG* 29.3(2000):282–294.

Warren, Karen. "Taking Empirical Data Seriously: An Ecofeminist Philosophical Perspective." *Ecofeminism: Women, Culture, Nature.* Bloomington and Indianapolis: Indiana UP, 1997. 3–20.

Wasserstrom, R. F. Wiles R Field Duty, U.S. Farm Workers and Pesticide Safety. Washington, DC: World Resources Institute, 1985.

Weaver, Jace. "Introduction: Notes From a Miner's Canary." *Defending Mother Earth: Native American Perspectives on Environmental Justice.* Ed. Jace Weaver. New York: Orbis Books. 1996. 1–28.

Weil and Kirchner. *Nature.* March 7, 2000.

Weisskopf, Michael. "Chemicals Used to Fight Termites Can Exact Human Toll." *Washington Post* 18 December 1987, final ed. First section A1.

Wernette, D. R., and L. A. Nieves. "Breathing Polluted Air." *EPA Journal* 18 (1992):16–17.

West, P. C., F. Fly, and R. Marans. "Minority Anglers and Toxic Fish Consumption: Evidence from a State-Wide Survey of Michigan." *Race and Incidence of Environmental Hazards: A Time for Discourse*. Ed. B. Bunyan and P. Hohai. Boulder: Westview Press, 1992.

"Where It All Began." http://www.environmentaldefense.org/documents/ 244_Where%All%Began.htm. (20 Sept. 2002)

"Who Are We, What We Do." MOMM: *Mothers Opposing Mandatory Medicine*. http://www.informedmomm.com/.

Wiant, Chris, "Risk Management at the Local Level." Commission on Risk Assessment and Risk Management. Symposium on a Public Health Approach to Environmental Health Risk Management. 8 August 1997, Washington, DC.

Williams, Florence. "Toxic Breastmilk?" *New York Times* magazine. January 9, 2005.

Wilson, Robin. "How Babies Alter Careers for Academics." *Chronicle of Higher Education* 50.15 (5 December 2003): 1–5.

Witherspoon, Nsedu Obot. "What About Our Children? An Ethical, Economic, Scientific and Political Discussion Around Children's Environmental Health." Children's Environmental Health Network. Talk. Siena College, Albany, New York. 8 April 2003.

Wolf, Jacqueline H. "Public Health Then and Now: Low Breastfeeding Rates and Public Health in the United States." *American Journal of Public Health* 93.12 (2003): 2000–2010.

Wollstonecraft, Mary. *A Vindication of the Rights of Woman: An Authoritative Text, Backgrounds, Criticism*. 1792. Ed. Carol H. Poston. New York: W.W. Norton, 1975.

Women in Europe for a Common Future (WEFC). Letters to European Commission, 30 November 2000.

"Women Warned of Chemical in Fatty Food." MSNBC News. 1 July 2003. http:www.stacks.msnbc.com/news/933424.asp (6 August 2003).

World Health Organization. "Global Data Bank on Breastfeeding." 13 March 2000.

———. *HIV and Infant Feeding: Framework for Priority Action*. Geneva, Switzerland, 2003.

———. "Press Release WHO/53." "Breastfeeding: A Community Responsibility." 6 1996. http://www.who.int/archives/inf-pr-1996/pr96-53.html.

———. "Principles for Evaluating Health Risks from Chemicals During Infancy and Early Childhood: The Need for a Special Approach." *Environmental Health Critieria* (1986): 59.

———. "Report of the Expert Consultation on the Optimal Duration of Exclusive Breastfeeding." Geneva, Switzerland. 28–30 March 2001.

Worobey, J. "Feeding Method and Motor Activity in 3-Month-Old Human Infants." *Perceptual and Motor Skills* 86 (1998): 883–1895.

Zahm S. H., and A. Blair. "Cancer Among Migrant and Seasonal Farm Workers an Epidemiologic Review and Research Agenda. *American Journal of Ind Med* 24 (1993): 753–766.

Zahm S. H., M. H. Ward, and A. Blair. "Pesticides and Cancer." *Occupational Medicine* 12.21 (1997): 260–289.

INDEX

Abbott Products, Abbot Laboratories, 179, 185

agricultural workers. *See* farm workers

Akwasasne, 142–143, 165

alar, 96

alarmism, 6, 30, 42 48, 84–85, 90, 97, 130

allergies, allergic reaction, 9, 95, 123, 146

American Academy of Pediatrics, AAP, 3, 9, 74, 80, 101, 108, 135–136

Apple, Rima, xii, 36–41, 110, 117–118

attention deficit disorder, ADD, 59, 95

Baby Friendly Hospital Initiative, 78, 108, 182

Baby Milk Action, 53, 75, 76, 78, 82–85, 179

Baumslag, Naomi, 36, 41, 46, 85, 129

bioaccumulation, 12, 63, 87, 90, 91, 145, 160, 181

body burden, 4, 6, 7, 8, 127, 144, 146, 157, 171–172, 178

breastfeeding advocates, 2, 7, 14, 16, 45–47, 49–53, 66, 83, 85, 106, 111, 112, 114, 129–130, 136. *See also* individual groups

breastfeeding and paid work, 102, 113, 119, 120, 135

Bullard, Robert, 141, 142, 150, 164

cancer, 5, 9, 10, 12, 27, 68, 72–73, 84, 87, 88–89, 95, 103, 122–125, 128–129, 131, 143–144, 154, 160–161, 174–177, 184, 187; breast, 6, 10, 15, 99–100, 103, 125; cervical, 10; lung, 103; overian, 10; pancreatic, 160; stomach, 160; testicular, 145

Carson, Rachel, xii, 12, 36, 61–69, 83–88, 92–93, 122–123, 126, 138–139, 142, 153–155, 173

Carter, Pam, 66, 104, 109–111, 119, 128

Center for Children's Health, CCHEC Health, 172–175, 184

CEP, Community Exposure Project, 160

Chavis, Benjamin, 141

childcare workers, 103, 109, 113, 119–121

Children's Environmental Health Network, CEHN, 74, 174

collective decision-making, 13

Community Exposure Project, CEP, 160

Cook, Katsy, or Gudgy, 129, 131, 144–147, 164–165, 183

Cumulative Exposures and Subsistence Fishing Project, Cumulative Assessments, 160

DDT, dichloro-diphenyl-trichloroethane, 16, 63–65, 86, 88, 92, 126, 130, 145, 154, 156, 161, 173, 180

209

Department of Environmental Conservation, DEC, 143–144, 162
diabetes, 3, 9, 123, 146, 177
dioxins, 10, 12, 16, 70, 72, 90, 127, 161, 180

ecological impact of formula feeding, 83, 103, 125
endocrine disruption, 12, 85, 95
Environmental Justice Movement, 18, 64, 92–94, 131–132, 140–141, 153–159
Environmental Protection Agency, EPA, 12, 65, 68–69, 72, 74, 80, 90, 129, 141–143, 149–150, 155, 160–161, 165, 178, 187
environmental toxicity as violence inflicted on body ,102, 140–152, 157, 167–169, 171–176
Eskimo, 181. See also indigenous people
essentialism/constructivism debates, 107
exposure level, 127, 161

farm workers, agricultural workers, farmhands, 85, 87, 148, 152–154, 157
Federal Food Drug and Cosmetics Act, FFDA, 150
FDA, Food and Drug Administration, 12, 94
Feeding Infants and Toddler Study, FITS, 185
Feminist, feminism(s), 7–8, 11, 14–16, 20–22, 41, 48, 52, 55, 63, 73–75, 94, 99, 100–107, 109–123, 126, 128–136, 165, 171–172, 181–182
Firestone, Shulamith, 115, 116
First National People of Color Environmental Justice Leadership Summit, 141
fish, 15, 61–62, 70–73, 90–91, 93, 95, 139, 140–142, 144, 145, 147, 156–159, 161–162, 170–172, 175, 181, 184, 187; advisories, 90–91, 162, 172; fishing as activity, 145, 147, 158–160, 162, 172; subsistence fishing, 158–160

FITS, Feeding Infants and Toddler Study, 185
focusing event, 96
Food and Drug Administration, FDA, 12, 94
formula. See infant formula

gender difference, 104, 116, 134
Golden, Janet, xii, 19–21, 25–30, 32–33, 35–36, 110, 118, 178
Gibbs, Lois, 131
Greenpeace, 70–72

Harlem Health Promotion Center, 159
Haudenosaunee, 143, 145. See also Akwasasne, Mohawk
Hausman, Bernice L., 101–102, 106, 110–114, 128, 182
heptachlorepoxide, heptachlor, 87
HIV and transmission through breastmilk, 13, 19, 33, 50–57, 103, 112, 129–131, 179
Human Milk Banking Association of North America, HMBANA. See milk banks

IBFAN, International Breast Feeding Action Network, 75, 78, 82, 84, 85
immigrants, immigration, 14, 19, 29–30, 91, 120–121, 156, 160–161
immunity suppression, 2, 95
IMF, International Monitoring Fund, 151
inequitable distribution of waste, 15, 48–49, 65, 76, 92–93, 140–143, 156–157, 160, 164–165. See also waste
indigenous peoples, 14–15, 46, 49, 72, 87, 94, 132–133, 172, 183. See also individual groups
Infact Canada, 52, 54, 75, 77, 83
Infant formula, 75–81, 94, 98, 101, 114, 179, 181–182; and promotion, 2–4, 40, 43–46; and ecological impact, See ecological impact of formula use; and risks of use, 30, 51
Iroquois, 143. See also indigenous peoples
Interagency Working Group on Environmental Justice, 164

International Breast Feeding Action Network, IBFAN, 75, 78, 82, 84–85
International Code of Marketing of Breastmilk Substitutes, direct marketing of infant Foods, 54, 78; See also World Health Organization
International Labor Organization, ILO, 4
International Monitoring Fund, IMF, 151
Inuit, 18, 85, 90, 92, 94, 181. See also indigenous peoples

JAMA, Journal of the American Medical Association, 39, 51, 58, 120, 124, 177, 179
Jelliffe, Derrick B. and E.F. Patrice, 19–20, 117

La Duke, Winona, 142–144, 146–147, 163–165
La Leche League, La Leche League International, 6, 9, 55, 71–72, 75, 99, 106, 135, 183
Law, Jules, 105–106, 111–112, 115, 123–124, 128, 135
Layne, Linda, xii, 105, 110, 182
lesbian parenting, 115
Lorde, Audrie, 100, 133, 144, 165
losing breastmilk supply, 44

malathion, 88
Maloney, Carolyn, Congresswoman, 135
Michels, Dia, 36, 41, 46, 85, 129
milk banks, milk banking, 20, 33–34, 57, 178, 180
miscarriage, pregnancy loss, 22, 59, 105, 110, 145, 182
Mohawk, 15, 92–93, 131, 143, 144–147, 164–165. See also indigenous people
Mothering, 9, 15, 38, 99, 101, 111, 115, 118
Mothering, 9, 99, 101
Mother's Milk Project, 193, 144, 157, 164

NAFTA, North American Free Trade Agreement, 151
National Institute of Environmental Health Sciences, NIEHS, 159
National Organization for Women, NOW, 101, 134–136, 141, 183
Native American Peoples. See indigenous people; see also individual groups
nativism, 30
Natural Resources Defense Council, NRDC, 86
Nestle, 38, 54, 179; Nestle Boycott, 9, 46, 67, 77–80
New York Times, 65, 67, 76, 79–80, 86–92, 144, 172, 174, 180–182
NIEHS, National Institute of Environmental Health Sciences, 159
NOW. See National Organization for Women
North American Free Trade Agreement, NAFTA, 151
North Carolina Breastmilk and Formula Project, 80

obesity, 9, 105, 123–124, 146

Palmer, Gabrielle, xii, 19, 21–24, 26–27, 30, 36–37, 40–41, 43–46
Patriot Act, 150
participatory action research, 136, 158
PBDEs, polybrominated diphenyl ethers, flame retardants, 4–5, 59, 92, 97–98, 126, 172–197
PCBs, polychlorinated biphenyls, 12, 16, 18, 65–67, 82, 86–89, 92, 127–128, 141–143, 145–146, 173, 180–181, 187
persistent organochlorine pollutants, POPs, 72, 86, 87, 94
pesticides, 10, 70–73, 85, 88–89, 93, 95, 129–130, 147–155, 157, 174, 181–184, 187; lack of handling instructions in English, 88; dumping in poor countries, 88
pesticide incident monitoring system, PIMS, 150
Pharr, Suzanne, 100

POPs, persistent organochlorine pollutants, 72, 86, 87, 94
policy, 8, 11, 16, 28, 46, 50, 63–64, 70–71, 80, 87–88, 94–95, 102, 130; abortion, 50; breastfeeding, 9, 75, 94–95, 102, 128, 135–136, 183; environmental, 18, 74–75, 82, 87–88, 140, 150, 159, 162, 165, 168–169, 174–179; gender-neutral, 134–135; Family Medical Leave Act, maternity leave, parental leave in academia, 6, 98,108; milk banking, 57; New Mothers' Breastfeeding Promotion and Protection Bill, 135; Pregnancy Discrimination Act, 134; public health, 105; Superfund, 141–142, 164, 171, 180. See also individual environmental policies
polybrominated diphenyl ethers, 4–5, 59, 92, 97–98, 126, 172–197
precautionary principle, 6, 98
pregnancy loss. See miscarriage

risk/benefit analysis, 43–44, 53–56, 64, 66–67, 69–70, 72, 83–84, 89–90, 111, 128, 130, 158, 160, 173–175
Ross Laboratories, Abbott Products Division, 185

safe exposure level. See exposure level
scare stories, 82, 84
Schector study, 69, 72, 90
"scientific childcare," "scientific mothering," 118
sexual division of labor, 103, 105, 115
Shoshone, 143. See also indigenous groups
Sierra Club, 49–50, 70–74, 87
Steingraber, Sandra, XII, 6, 7, 9, 11, 57, 66, 89, 95, 101, 103, 110, 122–129, 149, 150–151, 177, 182, 185
Superfund, 141–142, 164, 171, 180

Swanston, Samara, 160

Toxic Substances Control Act, 143, 180
Toxic Release Inventory, 130

Unborn Victims of Violence Act, 57–59
UNICEF, United Nations Children's Fund, 78, 109, 125, 186
United Church of Christ Commission for Racial Justice, 92
United Farm Workers of America, 148
United Nations Children's Fund, UNICEF, 78, 109, 125, 186
uranium, 93, 145
U.S.A. Patriot Act. See Patriot Act
U.S. Department of Health and Human Services, USDHHS, 1, 58

Violence, affecting milk quality, 28; associated with HIV status, 52; associated with slavery, 34–35; feminist attention to, 102, 121, 133. See also environmental toxicity as violence inflicted on the body; Unborn Victims of Violence Act

Warren County, North Carolina, 141
waste, 5, 10, 18, 42, 83, 125–126, 139, 181. See also inequitable distribution of waste
West Harlem Environmental Action, WE ACT, 159
WIC, Women, Infants, Children, 109, 136
World Bank, 151
World Health Organization, WHO, 9, 54–55, 57, 65–66, 68, 78, 85, 92, 185
WWF, World Wide Fund for Nature, 84, 85
Wyeth, 54, 78, 179